MILES/KILOMETRES

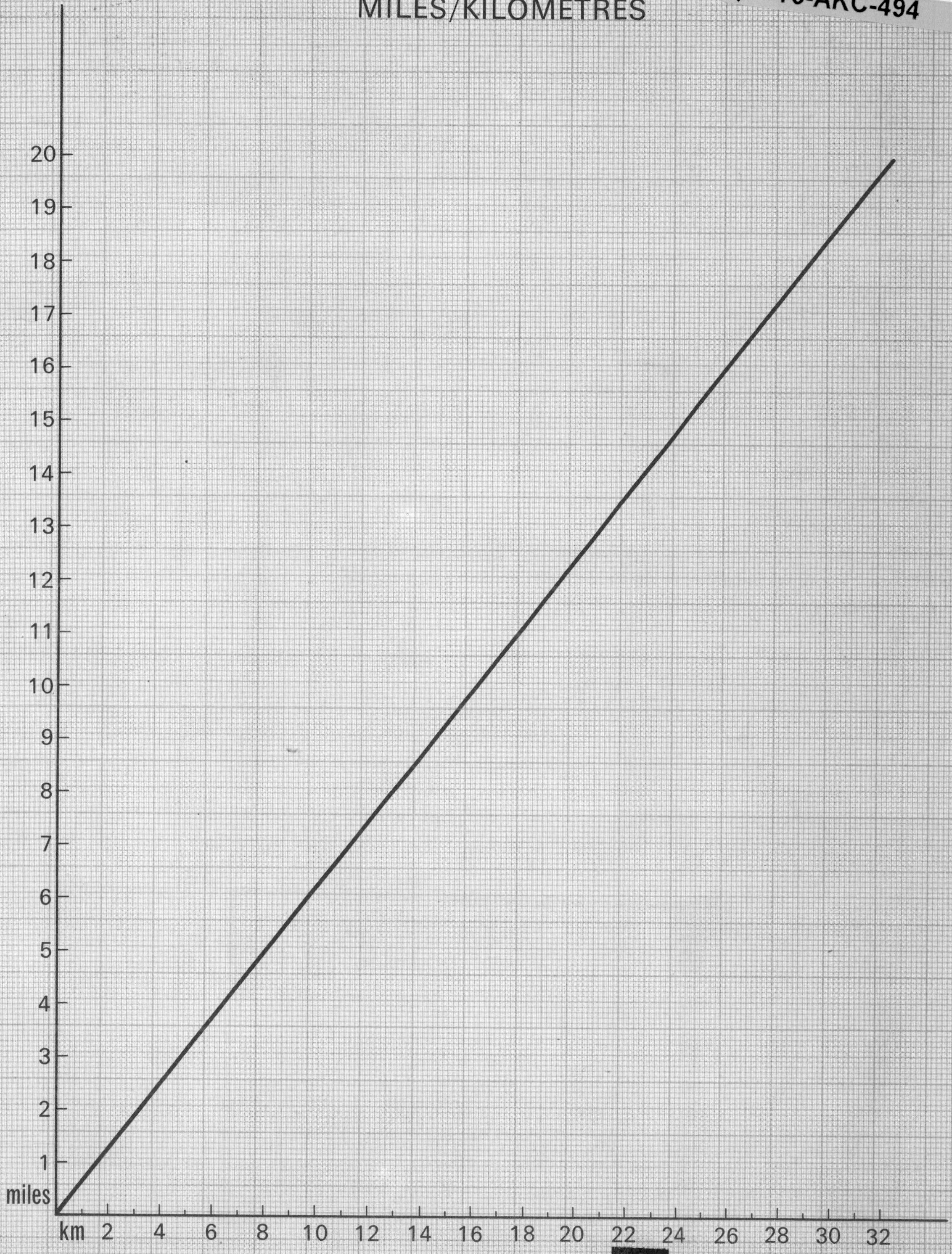

LANDSCAPE STUDIES -

an introduction to geomorphology

LANDSCAPE STUDIES -
an introduction to geomorphology

K. E. Sawyer

Senior Lecturer in Geography
Weymouth College of Education

Edward Arnold (Publishers) Ltd

First published 1970

SBN: 7131 1582 3

Printed by offset in Great Britain by
William Clowes and Sons, Limited, London and Beccles

PREFACE

Teachers of Sixth Form Geography have become aware during recent years of changing patterns of thought among University geographers. In the field of Geomorphology this has been evidenced in a questioning of the validity of W. M. Davis' Cycle of Erosion concept, an emphasis upon a study of the processes of landscape formation and an insistence on precise quantitative measurement as a vital part of Field Work. These changes are to some extent reflected in the syllabuses and questions for the Advanced Level examinations, and in particular in the insistence of many Examining Boards that the candidates shall show evidence of a close study of actual landforms.

The author believes that this book is in sympathy with these trends in that it attempts to lead the student from a careful study of selected landscapes to the consideration of general principles which arise from each example. Sections A to D of this book are primarily concerned with Processes, and in Section E the emphasis is on Structure, though the two must always be to some extent associated when an actual landscape is being studied. The conception of this book and the methods used are based upon the author's experience of teaching Geomorphology to Sixth Forms and his conviction that map and photographic interpretation must be integrated with a study of geomorphological concepts if the latter are not to become meaningless theories divorced from actual landscapes.

Throughout this book the word 'Landscape' is used to refer only to the physical landscape of an area. Many teachers will, of course, wish to broaden students' studies by relating the geomorphological aspects of the areas mentioned in the book to features resulting from man's occupancy of the land. Ideally each type of landscape should be studied in the field: not just looked at, but accurately observed, measured, recorded and analysed. In practice the opportunities for Field Work are limited and maps and/or photographs must be used to present unfamiliar landforms to the student. Each Study in this book seeks to apply the methods of the best field work to map and photograph study. Observation and recording are encouraged by the introduction of a variety of diagrammatic and statistical techniques. Description is the next stage, making use of the accurately observed facts and presenting them in a succinct verbal form. Only finally is the idea of interpretation introduced and the tentative nature of many explanations is made clear. By the application of such methods any landscape becomes a potential field for study and the student can come to appreciate that geomorphology is not limited to a relatively few 'classic' areas.

This book is written primarily for the Sixth Form student progressing from O Level (or equivalent) to A Level in Geography. The author also believes that, in its emphasis upon actual landscapes, much of the material will be of value to students of geography in Colleges of Education and in Further Education. At all these levels the student should be aware that no single text-book can provide all the material required. The Studies of this book should be regarded as an introduction to the topic concerned, and the Further Reading following each Study suggests sources from which the student may obtain greater depth and breadth in his or her studies.

In order to make the book available at reasonable cost, some detail has been removed from the map extracts while retaining all the necessary physical features and the character of the originals. In the interests of clarity with two-colour printing, the Grid Lines are not always marked on the map, but these can easily be pencilled in from the marginal data given. Students should have access to many of the Ordnance Survey sheets, details of which are given at the head of relevant Studies, and they will then be able to appreciate the human response in the areas studied and will be better equipped to relate these areas to their surroundings.

I must acknowledge my indebtedness to students of Weymouth College of Education and of West Bridgford Grammar School, Nottingham, whose ideas I have often used, and whose difficulties I have had in mind. I am also grateful to Dr. R. J. Small of the Department of Geography, University of Southampton, for reading the manuscript and for making many valuable suggestions, and whose article in 'Geography' (Volume 51, Part I, January 1966) provided the stimulus for writing this book.

K. E. S.

METRICATION

By 1975 it is likely that all Examination Boards will be setting Advanced Level papers in Metric (SI) units as part of the national policy of metrication by that date. It is with this target date in mind that it has been decided to use metric units in this book wherever possible, particularly as the author and publishers believe that schools will increasingly 'go metric' well before the target date in many other subjects in order to prepare pupils thoroughly for the new system.

In one respect the use of metric units in this book poses few problems since, with rare exceptions, only the units of length are required. The kilometre is familiar to students as the side of a grid square on all Ordnance Survey maps, and the approximation 10 feet = 3 metres will suffice for most purposes. To assist conversion, charts are printed as front and back end-papers.

From another aspect, metrication of this book raises difficulties since most of the maps used are, and will be for some years to come, in Imperial units. The policy has therefore been adopted that Imperial units (miles, feet, etc.) are given in addition wherever required for direct reference to a map.

CONTENTS

SECTION D DENUDATION UNDER ARID AND SEMI-ARID CONDITIONS

SECTION E LANDSCAPES DOMINATED BY ROCK TYPE AND STRUCTURE

BIBLIOGRAPHY

The following list is a full description of books to which reference is most frequently made in the reading lists following each Study. In the reading lists, where a reference is marked with an asterisk, the passage concerned either deals with the subject under immediate consideration at some length, or includes either vocabulary or concepts of a somewhat more advanced nature.

M. S. Anderson, *Splendour of Earth*, George Philip & Son Ltd., 1954
C. C. Carter, *Land Forms and Life*, Christophers (Publishers) Ltd., London 1961
G. Dury, *The Face of the Earth*, Penguin Books Ltd., London 1966
R. K. Gresswell, *Rivers and Valleys*, Hulton Educational Publications Ltd., London 1964
——: *Beaches and Coastlines*, Hulton Educational Publications Ltd., London 1957
——: *Glaciers and Glaciation*, Hulton Educational Publications Ltd., London 1962
——: *Geology for Geographers*, Hulton Educational Publications Ltd., London 1963
——: *Physical Geography*, Longmans, Green & Co. Ltd., London 1967
A. V. Hardy & F. J. Monkhouse: *Physical Landscape in Pictures*, Cambridge University Press, 1960
A. Holmes: *Principles of Physical Geology*, Thomas Nelson & Sons Ltd., 1965
N. K. Horrocks: *Physical Geography and Climatology*, Longmans, Green & Co. Ltd., London 1962
F. S. Monkhouse: *Principles of Physical Geography*, University of London Press Ltd., 1960
J. L. Scovel: *Atlas of Landforms*, John Wiley & Sons, 1966
B. W. Sparks: *Geomorphology*, Longmans, Green & Co. Ltd., 1963
J. A. Steers: *Coastline of England and Wales*, Cambridge University Press, 1964
——: *Coastline of England and Wales in Pictures*, Cambridge University Press, 1960
——: *Sea Coast*, Collins, 1963
A. N. Strahler: *Physical Geography*, John Wiley & Sons, N.Y. 1960
A. E. Trueman: *Geology and Scenery*, Penguin Books Ltd., London 1963
S. W. Wooldridge & R. S. Morgan: *An Outline of Geomorphology*, Longmans, Green & Co. Ltd., London 1959

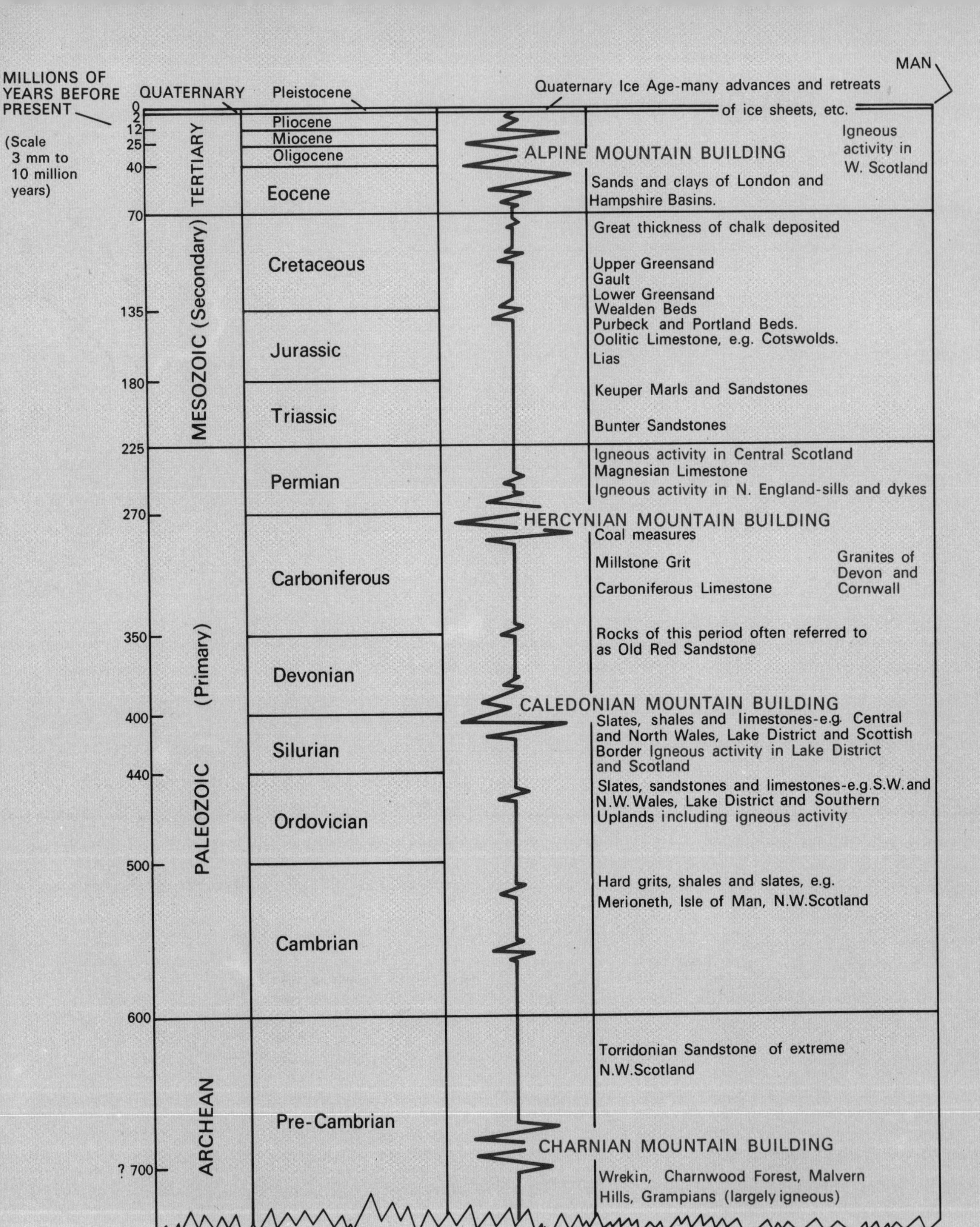

Fig 1 Geological Time Scale

Section A: Denudation in Humid Climates

Fig 2 Beacon Hill, Charnwood Forest—view N over the Midland Plain

Beacon Hill, Charnwood Forest, Leicestershire

Study A1

WEATHERING

O.S. 1 : 63 360 (1.6 cm to 1 km *or* 1 in to 1 mile)
Sheet 121, G.R. SK/509148

At 245 m (spot height 810 ft) above sea-level, 5 km SW of Loughborough, one may stand on rocks which are among the oldest to be found in the British Isles. The normal geological time scale (*Fig 1*) takes the story back to the dawn of life on Earth; to the Cambrian period which began 600 million years ago. Beacon Hill rocks belong to the pre-Cambrian Age, and by modern methods can be dated to some 700 million years in age. The difficulty of conceiving of such vast periods of time can be a fundamental obstacle to an understanding of the origins of landscape. Sudden changes do occur, as when a volcano erupts, or a major earthquake heaves billions of tonnes of rock by a few metres, or when a major cliff fall occurs along our coastline. But the important processes in **geomorphology** ('the making of the landscape') are extremely slow. Unless the vast time scale of the Earth's history is in some degree comprehended, it is very hard to imagine how these processes can have resulted in the extensive changes to which our landscape has been subjected.

Near the top of Beacon Hill, a slope strewn with angular slabs of rock of all sizes leads to the summit, from which emerge sharp edged masses of rock, predominantly dark greyish-brown in colour, with jagged edges and often fretted by narrow grooves. Where the slope of the hillside is less steep and exposed, a thin covering of gritty soil supports a vegetation cover predominantly of grass and bracken. From the foot of the hill, some 140 m below (i.e. below the 350 ft contour) extends a very different kind of environment, as the gentle undulations

Fig 3 Structure of an Inlier

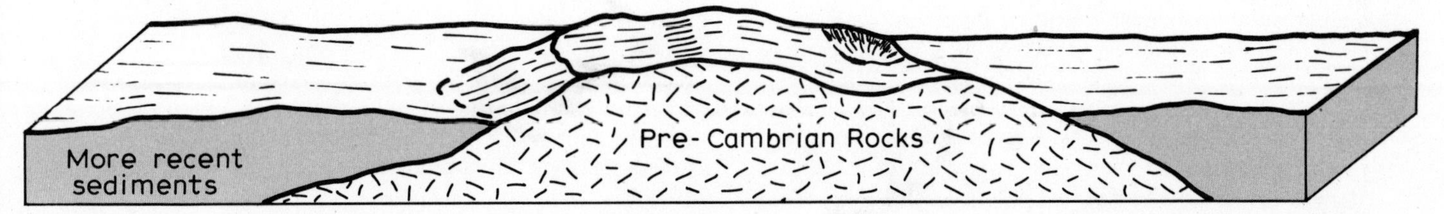

of the Keuper Marl of the East Midlands plain stretch beyond Loughborough, closely divided into hedged fields and mostly under pasture. Five and a quarter kilometres due E (G.R. 563150) can be seen the low, wooded and somewhat conical Hawcliff Hill.

Charnwood Forest, of which Beacon Hill is a part, is an 'island' of very ancient rock jutting out of much more recent rocks; a structure termed an **inlier** (*Fig 3*). It is likely that the whole of Beacon Hill was at one time buried beneath later deposits and has been exposed as this cover has been slowly removed by **denudation**. The term denudation is applied to the whole group of processes by which rocks are weakened and destroyed and the fragments removed, usually accumulating beneath the sea and eventually forming new sedimentary rocks.

The first stage in denudation must be the weakening and breaking of the solid rock. The magnitude of this task becomes apparent if the Beacon Hill rock masses are attacked with a geological hammer. These rocks, compressed and hardened by earth movements extending over hundreds of millions of years, are very resistant to such physical attack, yet the rock fragments scattering the ground leave little room for doubt that rock destruction is taking place actively under present conditions. The main agent of this destruction is in fact the weather, hence the use of the term **weathering.**

Mechanical weathering in the British climate is largely the result of **frost action.** Rain water collects in cracks and holes in the rock, and in winter becomes frozen. In changing into ice it expands by more than 9% of its volume and exerts great pressure (140 kg/cm^2). The crack is thus forced open as if by a wedge. The ice may melt by day, more water collects to fill the enlarged crack and then freezes at night. So day by day the crack, and many others like it, are enlarged until a piece of rock is broken off and falls (*Fig 4*). On steep, rocky mountain sides vast quantities of such frost shattered debris accumulate at the foot of the slope, forming **screes**, as along the E shore of Wastwater in the Lake District (*Fig 5*). Whilst significant frost action occurs under present-day climatic conditions (*see Further Study, Practical, 3, p. 4*) much of the accumulation of vast scree slopes in our mountain areas took place during the period of intensely cold climate immediately following the Ice Age of the Quaternary period.

Fig 5 Aerial view of Wastwater showing scree slopes

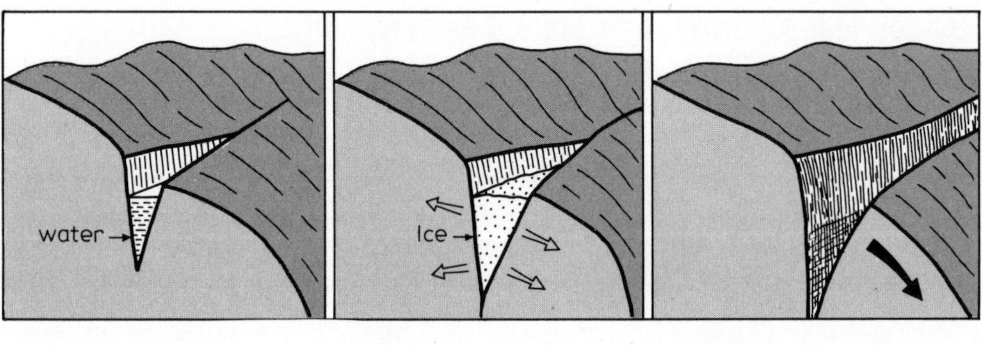

Fig 4 Frost action

(a) Crack in rock fills with water

(*b*) Water freezes and expands pushing rocks apart

(c) Repetition of process prizes off rock fragment which falls

In tropical climates frost is rarely, if ever, experienced. But in many tropical areas during a long dry season the difference between day and night temperature is extreme, and the sun beating down on rocks by day may raise their temperature to some 40°C above their night temperature. This will clearly lead to considerable expansion and contraction but these changes are not uniform: surface layers will be heated and expanded more than lower layers: dark particles will absorb more heat than lighter ones, different crystals in a crystalline rock will have different coefficients of expansion. All this will set up stresses within the rock which may lead to complete disintegration or to the peeling of layers to form rocks with rounded shapes, since corners and protuberances are the first parts to be affected. This removal of thin layers of rock is **exfoliation** ('the removal of leaves' (*Fig 6*).

Fig 6 Exfoliation

However important the weathering due to such diurnal temperature changes, laboratory research has shown that, except in the case of frost action, mechanical weathering is ineffective in breaking rock which has not first been weakened by **chemical weathering.**

When rocks are exposed to water, to atmospheric oxygen and carbon dioxide and to acids derived from plant decay, a number of chemical changes occur which combine to destroy the coherence of the rock. These processes collectively produce 'rock rotting' as opposed to 'rock breaking' which is the end-product of mechanical weathering. Evidence of chemical rotting is easy to find in urban areas where the acids of atmospheric pollution cause serious damage to stone carvings and to the face of stone buildings, but similar processes are at work where-ever rocks are exposed.

A very few rocks contain substances, notably salt, which are soluble in pure water, but much more common are rocks which become soluble when affected by rain water. In its passage through the atmosphere rain water absorbs some carbon dioxide and is thus a very dilute acid:

$$\underset{\text{Water}}{H_2O} + \underset{\text{Carbon dioxide}}{CO_2} \longrightarrow \underset{\text{Carbonic acid}}{H_2CO_3}$$

When this falls upon a limestone, the normally insoluble calcium carbonate of which the rock is largely made is converted into calcium bicarbonate which is soluble and is therefore washed away:

$$\underset{\text{Insoluble calcium carbonate}}{CaCO_3} + \underset{\text{Carbonic acid}}{H_2CO_3} \rightleftharpoons \underset{\text{Soluble calcium bicarbonate}}{Ca(HCO_3)_2}$$

This reaction is reversible as can be seen when 'fur' is deposited inside a kettle or when water, bearing calcium bicarbonate, drips into a cave and is evaporated to leave behind stalactites and stalagmites.

Recent research has shown that the greatest **solution** of limestone occurs within the first second of rain water striking the rock, and that after this water standing on limestone has little effect. The initial rapid solution produces the small vertical troughs divided by sharp ridges known as **lapiés**. It is suggested that the somewhat more rounded forms typical of limestone pavements on the Pennines (*see p. 121*) are the result of slower weathering beneath a past vegetation cover.

Calcium sulphate in its anhydrous state may take up water thus changing to a crystalline form which is gypsum. Such a change is referred to as **hydration** and involves a change in volume which may shatter adjacent rocks.

Many rocks contain iron in one form or another. These iron compounds react with oxygen in the presence of water to form iron oxides. Just as rust is a weaker substance than pure iron, so the end-products of this oxidation are weaker than the original forms of iron. Hence a rock affected in this way loses coherence and may be easily shattered by mechanical processes.

The end-products of rock rotting in temperate lands are the chemically stable substances of sand (silica SiO_2) and clays, which are more complex—e.g. Kaolinite ($Al_2O_32SiO_22H_2O$). In the higher temperatures of tropical regions, sands and clays are much less stable and are in turn replaced by bauxite (compounds of alumina, oxygen and water) and laterite (compounds of iron, oxygen and water). Under the heavy rainfall of some tropical lands chemical weathering may continue at depths of 15 m or more below the surface thus producing an enormously thick mantle of the residues of chemical weathering.

The chemical changes outlined above are but a few of the simpler processes involved in rock rotting. They illustrate the broad nature of the reactions involved and show that chemical weathering destroys rocks in three ways:

(a) by weakening the coherence between rock particles

(b) by the formation of soluble end-products which may be washed out by the rain, and

(c) by the formation of substances of different volume to the original so that stresses occur within the rock.

While not strictly 'weathering', it is convenient at this stage to consider the ways in which plants and animals contribute to the disintegration of rocks. This is often referred to as **biological weathering**. Plant roots may appear fragile in comparison with rock. But small rootlets find their way into joints and in growing they force the rock apart with incredible strength—the cellulose of which root walls are made is in fact stronger than metals. The presence of a vegetation cover also adds carbon dioxide, nitric acid, ammonia and other chemicals to the soil water, thus increasing its weathering capabilities.

Burrowing animals (rabbits, badgers, etc.) do not attack solid rock, but by bringing partially weathered rock fragments to the surface they accelerate rock destruction. The same is true of earth worms of which an average acre may contain 150 000, capable of raising ten to fifteen tonnes of finely divided material to the surface each year.

In all types of rock there are points of weakness which are more readily attacked by the weathering processes. This is most commonly seen in the case of **joints**, which are fracture lines normally at right angles to the **bedding** or layering of the rock. Since the joints are thus in the nature of vertical 'cracks' they offer a point of entry for

Fig 7 Close up of joint weathering on Charnwood

water and it is along the joints that weathering is concentrated, as may be seen from the close-up view of the rock outcrop at Beacon Hill. (*Fig 7. See also section on limestone pavements p. 121*). Some examples of exfoliation may be due to the formation of 'pressure (or load) release joints'. Rocks, particularly plutonic rocks, which were formed at considerable depths were inevitably subjected to great pressure from the overlaying rocks. As the latter were eroded the pressure was released and expansion occurred forming pressure release joints. A similar process can occur in older sedimentary rocks subjected to pressure by the accumulation of overlaying sediments of a later date which have been subsequently removed by denudation.

FURTHER STUDY

Practical

(1) A collection of rocks should be started, beginning with specimens of the home area and gradually extending this to include rocks from places visited during field study or on holiday. Label each specimen with its exact location.

(2) Whenever possible, attempt to collect both weathered and unweathered samples of a rock. Keep a record of the relative resistance of rock to mechanical attack, and of its ability to absorb water.

(3) Study any quarry face, cliff or similar exposure of rock, as well as stone buildings, and carefully note the effects of weathering. After a period of frosty weather, look for fragments of rock at the foot of stony walls or cliffs as evidence of rock shattering by frost.

(4) Students will find it useful to keep their own 'vocabulary' of terms used in Geomorphology, and it is suggested that all terms printed in bold type, e.g. **hydration**, should be entered in a notebook and a short definition added in the student's own words. Useful references for such work are: *The Penguin Dictionary of Geography* by W. G. Moore and *A Glossary of Geographical Terms* by L. D. Stamp (Longmans).

Reading

Dury: *The Face of the Earth*, Ch. 2, pp. 4–10
Greswell: *Rivers and Valleys*, Ch. 2, pp. 13–22
——: *Geology for Geographers*, Ch. 8, pp. 90–102
——: *Physical Geography*, Ch. 12, pp. 99–111
Hardy & Monkhouse: *Physical Landscape in Pictures*, p. 29
*Holmes: *Principles of Physical Geology*, Ch. III, pp. 36–63; Ch. VII, pp. 142–263; Ch. IV, pp. 386–410
Horrocks: *Physical Geography and Climatology*, Ch. 1, pp. 4–6
Monkhouse: *Principles of Physical Geography*, Ch. 4, pp. 77–81
——: *Landscape from the Air*, p. 13
*Sparks: *Geomorphology*, Ch. 3, pp. 22–42
Strahler: *Physical Geography*, Ch. 22, pp. 311–318
Wooldridge & Morgan: *An Outline of Geomorphology*, Ch. XI, pp. 133–137

Maiden Castle, near Dorchester, Dorset

SOIL CREEP (MASS MOVEMENT)

O.S. 1:63 360 (1.6 cm to 1 km *or* 1 in to 1 mile)
Sheet 178, G.R. 670885

Huge earth moving feats such as those involved in the construction of Maiden Castle, just as the building of modern motorways or dams, serve to remind us that Man is an agent in the formation of the landscape. Maiden Castle consists today of a series of grass-covered ramparts, at least three and in places more, encircling the summit of a chalk hill 3.2 km S of Dorchester. These banks rise some 15 m above the intervening ditches and are a most impressive reminder of the prehistory of the area. Yet archaeological investigation shows that the ramparts were once much higher and steeper, and the ditches correspondingly deeper. In the two thousand years since their construction, weathered and broken chalk has been steadily slipping from the banks into the ditches and evidence that this movement is still continuing may be seen in the minor corrugations running horizontally across the face of slopes. These are sometimes colloquially referred to as sheep-tracks, but in reality they are one indication that the residue of weathering is moving slowly down the slope in spite of a vegetation cover. These turf rolls or **terracettes** represent the surface expression of a gradual downhill movement of soil periodically lubricated by surface water and are to be seen on steep, grass covered slopes on a wide variety of bedrock. Other indications which demonstrate a similar downhill move-

Fig 8 Types of evidence which indicate soil creep

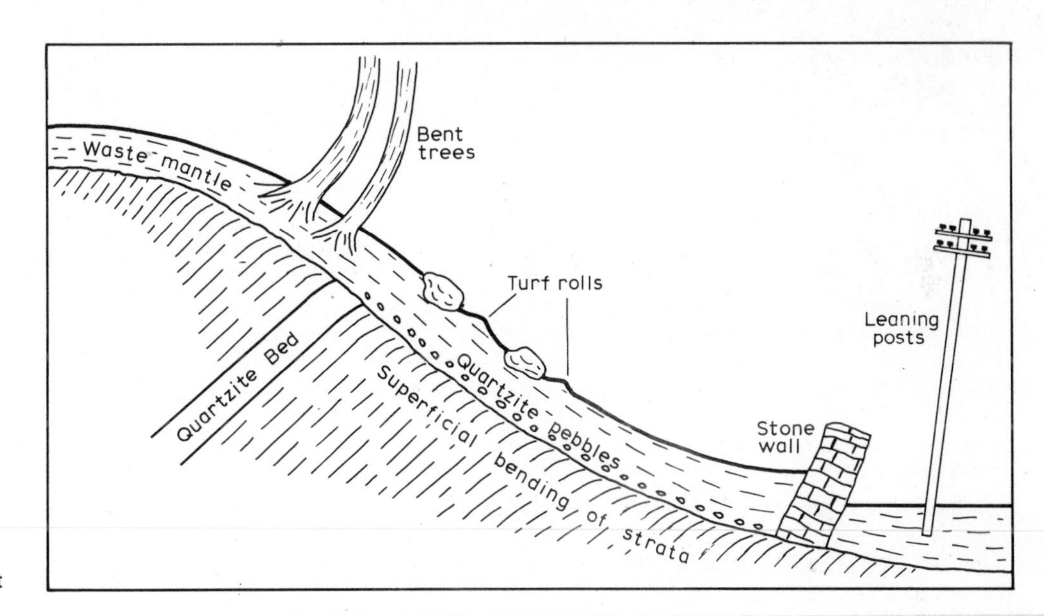

Fig 9 Maiden Castle, Dorset

Fig 10 A raindrop landing on wet soil produces a miniature crater spraying droplets outward

ment of soil are summarized in the diagram (*see Fig 8*). Trees growing on an unstable slope begin to lean outwards; then as the tree grows it corrects this lean so forming a curved trunk. The processes by which this downhill movement, **soil creep** or **mass movement** occur are many. Simple **slumping** following lubrication by heavy rain has been suggested above. When a large raindrop strikes wet earth, droplets are scattered, (*Fig 10*) and more droplets containing soil particles will fall on the downhill side.

When rock is affected by expansion and contraction, whether as a result of diurnal temperature changes (i.e. changes during 24 hours) or of freezing and thawing, there is a greater expansion in a downhill direction and a greater contraction on the uphill side, thus producing a nett downward movement.

In the above paragraphs it has been suggested that the downslope movement of soil and weathered rock is normally quite slow, and this is especially true when a close vegetation cover protects the soil and its roots bind the soil together. Under such circumstances on a moderate slope the loss by soil creep will be balanced by the progressive weathering of rock so that the land is slowly lowered but the soil thickness remains constant (*Fig 11*).

If the vegetation cover is removed, e.g. by ploughing, tree felling, over grazing, etc. this balance is destroyed. Under certain conditions of soil, slope and climate the removal of soil may be greatly accelerated to a rate when it exceeds the production of new soil by weathering. The resultant loss of soil, which may be almost complete in severe cases, is **soil erosion**, and this washing away of soil should not be confused with the quite normal soil creep which is an essential part of the process of denudation in humid lands, transferring the end-products of weathering to the foot of a valley whence it is removed by rivers.

FURTHER STUDY

Practical

(1) Seek out likely spots in your home area where soil creep may be expected. Sketch any available evidence that this process is occurring.

(2) Look carefully at your local soils. Describe them in terms of colour, texture, porosity, etc. (A detailed study of soil is beyond the scope of this book but suggestions for reading on this topic are given below.)

(3) A study of mass wastage must take into account the **angle of slope** of the ground. This measurement, which is perhaps more useful to the geomorphologist than 'gradient' can easily be measured with a home-made **clinometer.** (*Fig 12*) A piece of drinking straw, preferably the stronger plastic kind, is glued with Araldite or similar adhesive to the straight edge of a plastic protractor. A small hole is bored at the centre of the semi-circle and a weight attached by a thread. By siting on a pole, or person, of similar height to the observer the angle between the slope and the horizontal can be found by subtracting the smaller reading against the thread from 90°. Using this method the student can learn to appreciate differing degrees of slope, and to relate this to evidence of soil creep. At a later stage the value of slope mapping will be indicated.

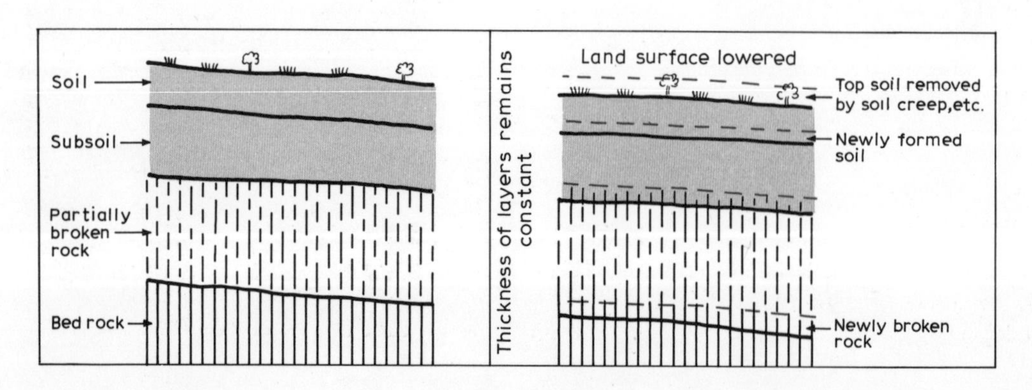

Fig 11 Creation of new soil as surface is slowly lowered. The removal of vegetation may speed erosion, removing top soil before new soil can form

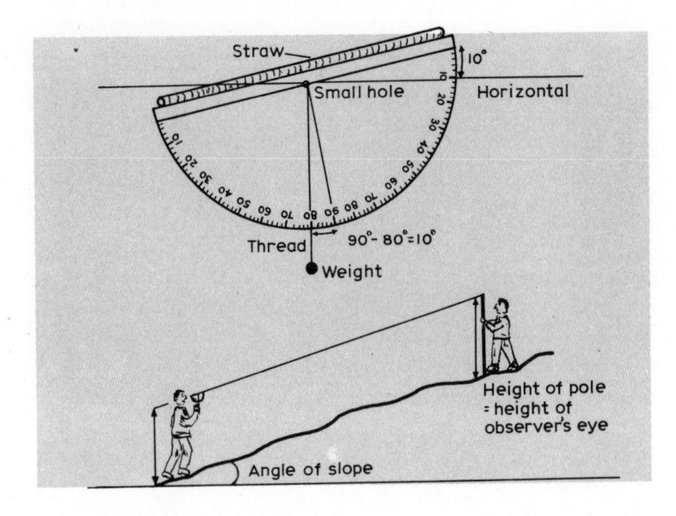

Fig 12 A simple clinometer

Reading

Dury: *The Face of the Earth*, Ch. 2, pp. 10–15
Gresswell: *Rivers and Valleys*, Ch. 2, pp. 24–26;
——: *Physical Geography*, Ch. 13, pp. 112–121; Ch. 14, pp. 122–130; Ch. 33, pp. 447–458
*Holmes: *Principles of Physical Geology*, Ch. XIV pp. 400–406; Ch. XVII pp. 468—472 and 481–502
Horrocks: *Physical Geography and Climatology*, Ch. 16, pp. 253–267
*Money: *Climate, Soils and Vegetation*, Univ. Tut. Press, 1965
Monkhouse: *Principles of Physical Geography*, Ch. 6, pp. 109–110 Ch. 18, pp. 366–387;
*Sparks: *Geomorphology*, Ch. 4, pp. 43–54
Strahler: *Physical Geography*, Ch. 16, pp. 236–242; Ch. 17, pp. 243–255; Ch. 22, pp. 318–325; Ch. 23, pp. 330–334

RIVERS—AN INTRODUCTORY STUDY

Study A3

Before considering examples of rivers in the landscape, it is useful to clarify ideas on how water flows.

The suggestions for practical work which follow are intended to stimulate thought and observation and can be modified according to local circumstances. First shape in sand or soil a mound or slope 50 cm to 60 cm in height—the author has made use of a school high jump pit for this work. Then water this using a hose or watering can with a fine rose. Students should observe carefully and note the changes which occur. Water will be seen to collect in hollows as the sand becomes saturated, and then begin to flow down the slope, taking the line of least resistance. When this happens, stop the 'rain' and try to decide why this course was followed by the 'river' in preference to alternatives. Was this the steepest slope? Was the mound less compact at this point? etc.

Now note accurately the course of the 'river' and start the 'rain' again for a short time. Has the river channel now changed? It is likely that bends in the course of the river may be noticeable. These may be related to some minor obstruction in the channel (a stone, etc.) but not necessarily so. Measure the **wavelength** of the curves produced and note any tendency for this to increase downstream. Similarly note whether the **amplitude** (*Fig 13*) of the curves increases (a) downstream, (b) when water is allowed to flow for a longer period.

Careful observation of the behaviour of the 'river' on the sand pile should reveal facts about the way in which water flows round curves. Where is the current strongest in relation to the curve? Where is the river bank eroded? Where is surplus sand deposited? The answers to these questions and their relevance to the landscape will be found in *Study A7* (*p. 18*).

Further experiment with the sand pile can reveal many other important aspects of the 'behaviour pattern' of

Fig 13 Wave length and amplitude of river meanders

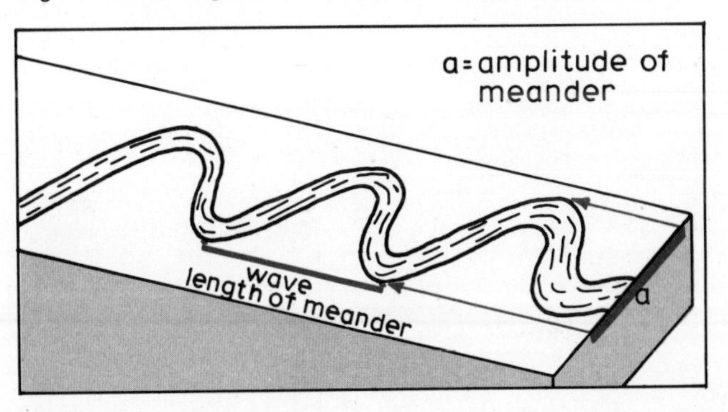

rivers. **Erosion**[1] by the 'river' is clearly seen to increase the depth of the valley, and eroded sand is **transported** along the river to be deposited at a lower level. When the river is checked at the base of the slope it becomes choked with its **load**; it may divide into numerous channels when it is said to be **braided.** By burying a thin rock slab just below the surface of the slope it is possible to show how one type of waterfall is caused.

In all the experiments conducted in this way it is important to remember that, useful as such practical demonstrations are, they are not complete replicas of natural conditions. On the sand pile, the 'river' is constantly losing water by percolation and is not gaining water from tributaries throughout its length; therefore the **volume** of our miniature stream decreases, rather than increases as it does in nature. Secondly, the speeding up of the time scale in our experiments removes the element of weathering on valley sides. Thirdly the whole action occurs on an unconsolidated mass rather than on solid rock of varying characteristics. Yet with all these limitations, much can be learnt and with ingenuity other demonstrations can be similarly arranged.

Before considering further facts about river flow it is necessary to define some of the terms used:

The **energy** of a river at any point depends on volume and velocity. Remember that up to 90% of the river's energy is used to overcome friction with its bed and banks. The remaining energy is available to transport a load of rock debris, the maximum weight of which on any stretch of the river represents its **capacity**.

The **load** may be transported in three ways: (a) in solution, (b) in suspension and (c) by rolling intermittently along the bed, a process known as saltation (the traction load). For a river of the size of the Mississippi the load is enormous:

ANNUAL LOAD OF MISSISSIPPI (*average*)

	Millions of tonnes
Solution load	200
Suspension load	500
Traction load	50
Total	750

It has been calculated that the chief rivers of the world together discharge about 8000 million tonnes of alluvium into the sea per annum, representing an average of 80 tonnes per square kilometre of their basins. This figure represents a lowering of the land surface by 30 cm in nine hundred years.

The **competence** of a river describes the weight of the largest particles which can be moved at any point. It may well be that a stretch of river is not loaded to capacity, yet the bed is littered by stones. A handful of sand

[1] **Erosion** is the general term used to describe the wearing away of land by processes where a definite movement is involved. Note the comparison with 'weathering' where the movement is very slight. Weathering and erosion together equal denudation—the sum of all processes by which the land surface is lowered.

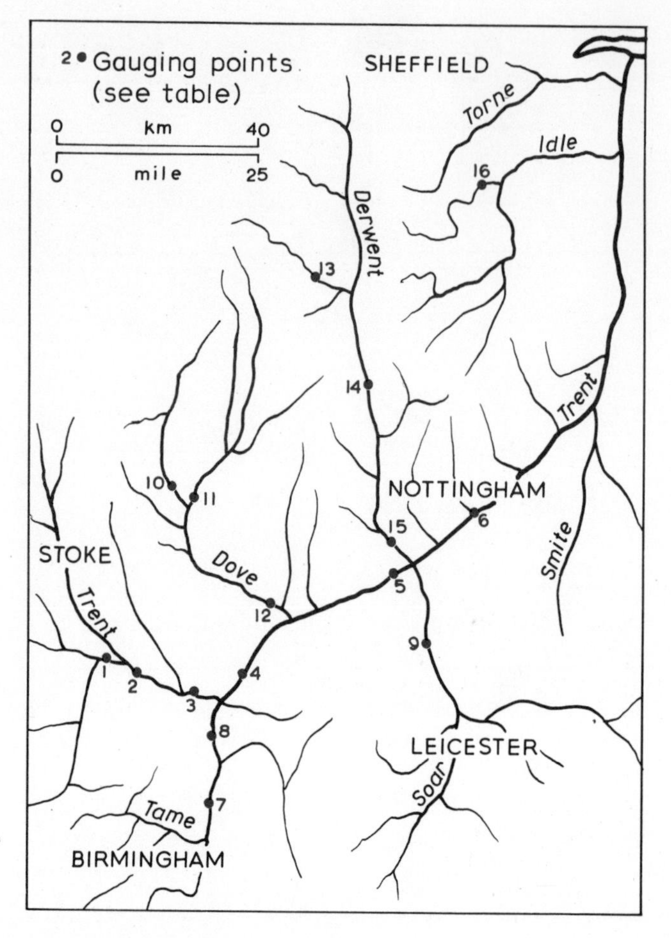

Fig 14 Rivers of the Trent Basin

SURFACE VELOCITY AND GRADIENT IN RIVERS OF THE TRENT BASIN

Gauging Point No.	Gradient of River 1 in . . .	Surface Velocity, metres per second (Autumn 1965)
1	1700	0·3
2	1700	0·6 to 0·9
3	1200	0·6
4	1200	0·6
5	1800	0·6 to 0·75
6	2100	0·75
7	950	0·6 to 0·9
8	1700	0·6
9	1300	0·08 to 0·15
10	450	0·6
11	400	0·45 to 0·6
12	650	0·9
13	250	0·45
14	500	1·2 to 1·5
15	775	0·75
16	1700	0·6

Reproduced with the permission of the Trent River Authority.

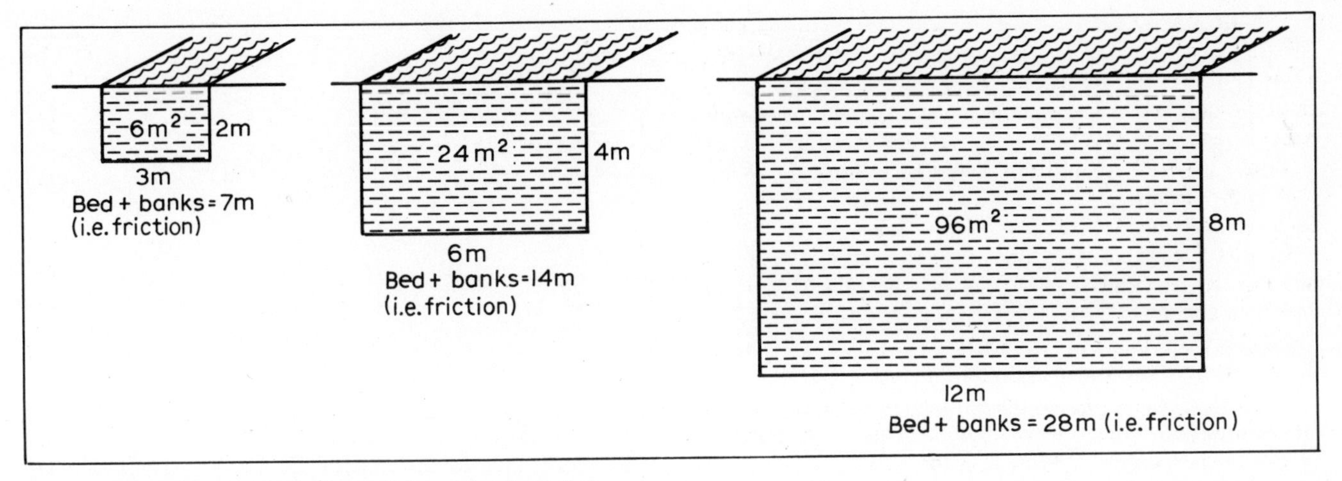

Fig 15 Comparative cross sections of rivers of various sizes

thrown into the water is at once carried away, yet the stones are not moved until an increase in velocity (as in a flood) increases the river's competence. This fact helps to explain the gradation in the nature of river bed deposits, decreasing in size of particles as the sea is approached.

It is commonly assumed that a river in its steep upper course is flowing more rapidly than in its lower course where the gradient is slight. An imaginary picture of a hustling babbling brook contrasted with a large, smooth river appears to support this. But actual measurements show that this is not so, and careful observation of the speed of flow in the centre of a major river will reveal that 8 km/h may be frequently recorded during spate, a rate which is rarely exceeded by our 'babbling brook'. Readings supplied by the Trent River Authority (*Fig 14*) demonstrate that surface velocity is not directly related to gradient. Why is this?

It has been stated above that a large percentage of a river's energy is used to overcome friction against the banks and bed. If the channel of a river remains constant in shape, then this friction becomes less for each unit of volume as the river increases in size (*Fig 15 above*). In the case of the largest river the index for friction (represented by the sum of the sides and bed of the channel, and often referred to as the **wetted perimeter**) is just over a quarter of the index for volume (the area of the cross-section), although in the smaller stream the friction index is larger than the volume index. It can be seen that a larger proportion of the small river's energy is used overcoming friction and the speed of the river is thus reduced.

A similar effect is shown in *Fig 16 below*, where two rivers of equal volume but differing cross-sections are shown. Friction is greater in the case of the shallow river, a fact which will reduce its velocity and hence its energy and capacity to transport a load.

The shape of the cross-section of a river valley similarly affects the flow of a river. This becomes of significance when flood water increases the volume of the stream as can be seen from the diagrammatic examples (*Fig 17*).

River A flows in a deep, steep-sided valley, so an increase in volume is accompanied by a reduction in relative friction and therefore an increase in river energy. So River A, in flood, will erode actively and its increased speed gives it the competence to move large boulders which will merely lay on the stream bed under normal conditions.

Fig 16 Comparative cross sections of rivers of equal volume but contrasted width and depth

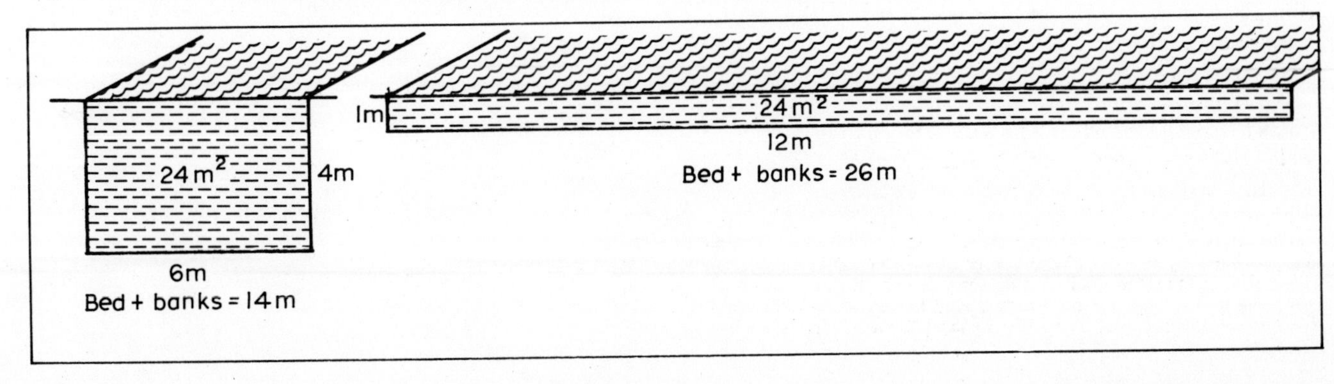

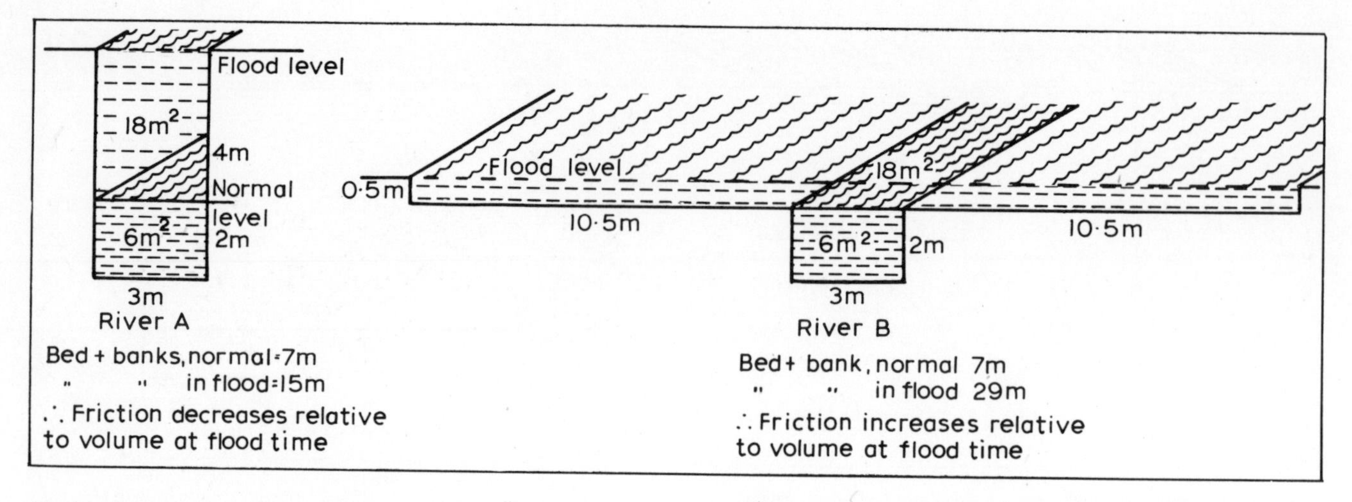

Fig 17 Cross sections of river channel showing effects of equal volume of flood water in a narrow valley and a broad valley

River B, by contrast, normally fills a bed cut into a broad flood plain. Flood water spreads out widely across this plain with a resultant increase in friction relative to volume. The consequent reduction in energy is most noticeable in the shallower flood water where deposition is likely to occur.

Both the extra erosion by River A and the deposition by River B which occurs at flood time are only very slowly modified when the river returns to its normal level and it is thus easy to understand the observed fact that the form of a river channel is adjusted to flood conditions, rather than to normal river levels.

Gradient, therefore, is only one of many factors which influence the velocity and hence the energy of a river. In the chapters which follow, some actual rivers will be examined in relation to the above facts and concepts, and their influence upon the landscape considered.

FURTHER STUDY

Practical

(1) As suggested in the text, many aspects of river behaviour may be examined on a sand pile.

(2) Where a river or stream is accessible, careful observations should be made to find the extent to which the results of the sand-pile experiments can be applied to real conditions. Any significant or puzzling features should be recorded by sketch-map, sketch or photograph.

(3) Measurements of the rate of flow of various streams may be made using either a colouring substance (e.g. potassium permanganate) or floating objects (e.g. weighted fishing float to reduce wind effects) timed over a measured distance. In the case of a large river it may be possible to record flow rates in the centre of the stream and close to the banks. Such recordings may be related to the gradient and the form of the channel. Ideally such readings would also be made at varying depths.

(4) Where a stream can be regularly visited it may be possible to graph changes in level and velocity through the year and relate these both to weather conditions and to changes to the river channel, e.g. bank erosion, deposition, movement of large boulders.

(5) The load of any stream can be assessed by collecting samples of water, and allowing these to stand (or filtering) to see the suspended load. It may be possible to analyse chemically at least some part of the solution load also.

Reading

Dury: *The Face of the Earth*, Ch. 7, pp. 75–80
Gresswell: *Physical Geography*, Ch. 16, pp. 146–152
Holmes: *Principles of Physical Geology*, Ch. XVIII, pp. 510–514
*Sparks: *Geomorphology*, Ch. 5, pp. 76–84
*Strahler: *Physical Geography*, Ch. 23, pp. 334–346

Wellhead Springs, Westbury, Wiltshire

HEADWARD EROSION

O.S. 1:63 360 (1.6 cm to 1 km *or* 1 in to 1 mile)
Sheet 166, G.R. ST/877502
O.S. 1:25 000 (4 cm to 1 km *or* 2½ in to 1 mile)
Sheet ST/85

The N facing scarp of Salisbury Plain rises steeply from 120 m to 180 m (i.e. 400 ft contour to 600 ft contour) about 1.5 km south of the town of Westbury, Wiltshire. A turf covered slope, locally known as Grassy Slope, forms a footpath down the scarp face, flattening out at the foot of the hill and leading between arable fields across part of a more or less level surface formed on the outcrop of Lower Chalk. At the N end of this track is a pumping station for Westbury's water supply, and immediately across an E–W lane is a steep drop to a narrow stream valley. At the base of the slope one source of this stream can be seen. From a small horizontal hole in the rock a **spring** emerges, though this produces only a very little water at certain seasons. Immediately above the spring the slope is almost vertical and shows clear signs of active soil creep though the top of the slope is less steep and the turf has a more stable appearance. From the spring, and joined from the E by another small rivulet, the stream flows NW through its steep-sided, narrow valley, eventually to join the R. Biss which flows into the Bristol Avon.

In considering the significance of the features described above, reference must first be made to the occurrence of springs. Rain water is, in part, returned directly to the atmosphere by evaporation; some of the water flows over the surface as **run off** and some is absorbed into the soil and into the underlying rock. The relative proportions depend on such factors as air temperature, degree of slope, vegetation cover and nature of the bed rock. In the case of chalk, as at Westbury, a high proportion enters the rock and collects, particularly in narrow fissures, moving steadily downwards until its progress is checked by impermeable strata. The lower levels of chalk are thus saturated with water, the upper surface of saturation being known as the **water table.** Where the water table reaches the surface at the edge of the hills, springs are likely to occur. Because the water is not able to move absolutely freely through the small spaces in the chalk, the water table tends to be at a higher level under the higher ground than where the spring occurs, thus producing a pressure gradient under the influence of which water moves steadily towards the spring, as shown by brown arrows on *Fig 18.*

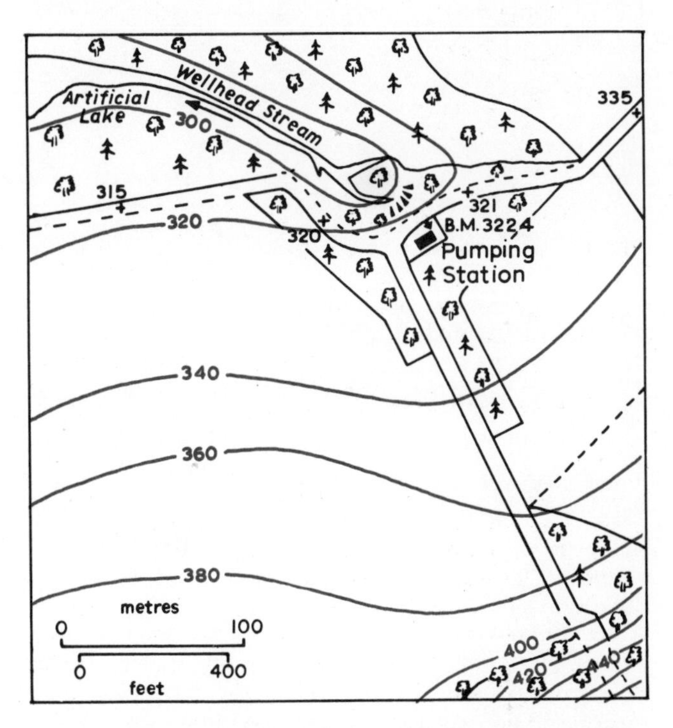

Fig 18 Wellhead Springs, Westbury, Wilts (contours in feet)

In the case of the edge of Salisbury Plain, the scarp is actually in two parts—one formed by the Upper and Middle Chalk, followed by the 1–2 km wide Upper Greensand and Lower Chalk Platform bordered by a second minor scarp (*Fig 20*). It might be expected therefore that springs would occur at A, but Wellhead Springs lie at the head of a sharp valley (or **re-entrant**) cut back into the

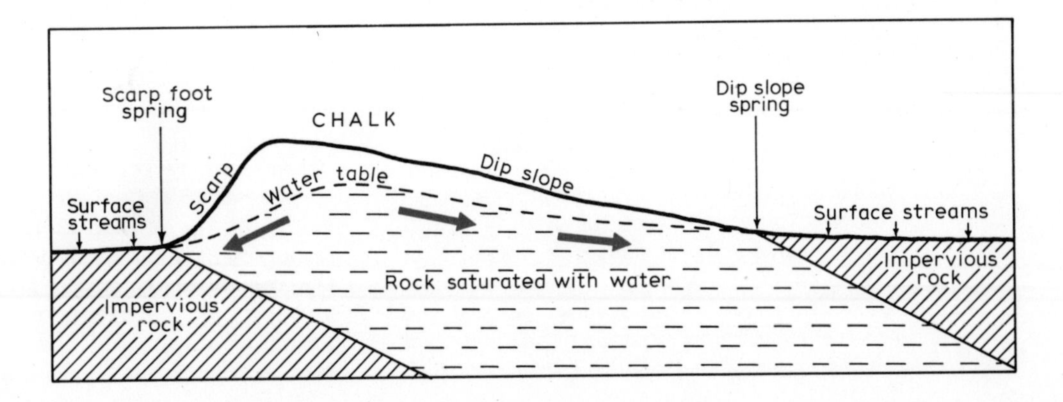

Fig 19 Generalized section to show relationship of water table, springs and rock formations in a chalk cuesta

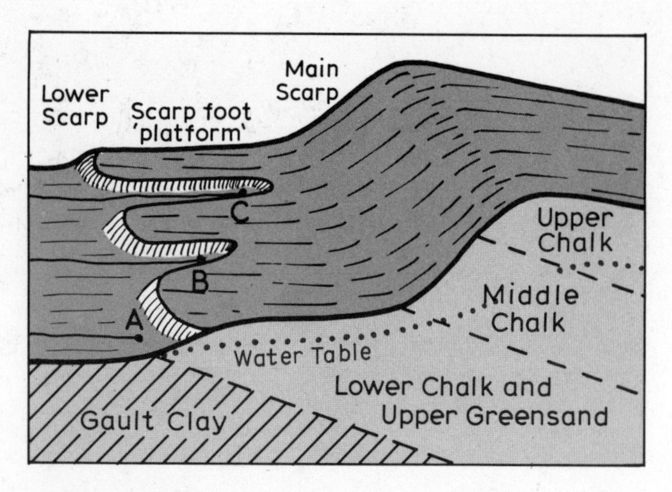

Fig 20 Spring recession across a Lower Chalk and Upper Greensand platform

lower chalk platform to B and C. The spring has thus retreated and the valley lengthened in a headward direction. Study of the present spring shows this process of **headward erosion,** or **spring-sapping** at work. Softened, saturated rock round the spring itself is steadily washed away downstream thus undercutting the slope behind the spring so that material slumps down into the stream and is washed away. This produces the very steep slope immediately above the spring. In addition rain water is concentrated by the semicircular form of the valley head and this helps to accelerate the effect of valley recession.

The fact of headward erosion, one example of which has been studied above, occurs to a greater or lesser degree at the source of almost every river, even when the river has its origin in the collection of run-off water rather than a spring. And the effect has a considerable significance in the study of river patterns, in rejuvenation and river capture. (*See Studies A10 to A13.*)

FURTHER STUDY

Practical

(1) Most maps of the spring-line zone at the foot of chalk scarps, especially on the 1:25 000 scale will show similar streamlets with valleys lengthened by headward erosion. These and any examples observed in the field should be sketched and described.

(2) Study any spring which is accessible. Make careful observations and draw an annotated sketch map and field sketches. Particularly notice any evidence of soil creep and/or headward erosion around the spring. Can you measure the volume of flow from the spring? Is there any indication that this flow varies from season to season, or that the spring has had a greater volume in the past.

(3) Read the article 'Underground Water' by J. A. Barrett, in *Geography*, Volume 51, p. 224, 1966. If possible try to put into practice some of the ideas contained in the article.

(4) After reference to some of the books cited below, prepare a series of diagrams to show some of the geological conditions which can give rise to the formation of springs.

(5) If you live in a district with wells and/or a waterworks try to find out the depth(s) from which water is obtained. This will give some indication of the depth of the permanent water table.

Reading

Gresswell: *Rivers and Valleys*, Ch. 3, pp. 27–32; Ch. 4, p. 39

——: *Physical Geography*, Ch. 15, pp. 138–140

*Holmes: *Principles of Physical Geology*, Ch. XV, pp. 411–434

Monkhouse: *Principles of Physical Geography*, Ch. 5, pp. 82–92.

*Strahler: *Physical Geography*, Ch. 21, pp. 298–310

*Wooldridge and Morgan: *An Outline of Geomorphology*, Ch. XVIII, pp. 246–253

The Processes of River Erosion

VERTICAL EROSION

Study A5

Half fill a jar with water and add some sand. The sand will quickly sink to the bottom, but if the jar is agitated briskly the moving water will sweep the sand up with it until the motion ceases and the sand sinks again. Now direct a jet of water from a hose on to a bank of earth. The force of the water will wash out loose material and carry it down the slope. These experiments demonstrate the power of water alone to erode and transport rock debris—powers referred to as **hydraulic action** when considered as part of the process of river erosion. The extent of the erosion and transportation in both our examples, as in a river, is determined by the velocity and the volume of the water. By these processes a river is clearly capable of deepening and widening its channel in unconsolidated rock debris. In more compact and resistant rocks the same processes operate, though much

Fig 21 The river Frome, Maiden Newton, Dorset

more slowly, to enlarge cracks or erode weak spots until pebbles or boulders are released to join the mobile load of the river. But other processes are also at work.

If careful observation of the surface velocity of a river is carried out, it will almost invariably be found that water is moving at a greater speed in the centre of the channel than close to the banks. Theoretically such a state can be achieved by an even fall in speed from the centre to the banks so that a smooth forward flow is maintained throughout. In practice, however, due to irregularities in the shape of the channel, it will usually be found that roughly circular whorls of water are set up and it is common to find places against the banks where the water is actually moving back upstream. This phenomenon is known as **turbulence** and is characteristic of

Fig 22 Stream flow, with horizontal turbulence

liquids flowing over an uneven surface. Turbulence also operates in a vertical plane (*Fig 23*), associated with the fact that surface velocity is greater than that along the stream bed. Vertical turbulence may be observed in a rapidly flowing stream in the form of ripples or wavelets produced by water swirling to the surface from the river bed. *Figure 21* shows this effect on a river of alternating deep and shallow stretches; where the river is shallow the velocity must increase in order to pass the same volume of water per hour as the deeper sections (since the width remains constant). This speeding up, perhaps combined with a more irregular rocky bed, produces turbulence which can be seen in the form of 'broken water'.

The significance of turbulence in the present context lies in its effects upon stream erosion. Firstly the velocity of water at certain points is increased with a resultant concentration of hydraulic action. Secondly, if the velocity reaches 12 metres per second as it may do in mountain torrents or in waterfalls a process known as **cavitation** occurs. Turbulence leads to marked pressure variations within the water and bubbles of water vapour form where pressure is low and implode (the opposite of explode) with considerable violence when carried to a point of higher pressure. The miniature shock waves set up by such implosion have a strong effect upon nearby rock, and even the hardest rock becomes pitted and finally broken by the continual repetition of this process.

Flowing water alone, therefore, by hydraulic action (including cavitation) produces rock debris, and immediately the river becomes thus armed with boulders, pebbles and sand grains a new group of erosive processes occur, collectively known as **corrasion**. The simplest expression of corrasion is the wearing away of the rocks of the stream channel by a constant bombardment by the rock fragments making up the stream load. Corners and irregular projections are the first part to be worn away and the eventual result is a smoothing of the rocks forming the stream bed. The top surfaces of the rocks of the temporarily dry stream bed shown in *Fig 26* illustrate this characteristic.

The effects of turbulence must also be considered in connection with corrasion. Rock fragments in the load are carried sharply downwards by vertical turbulence adding to their impact velocity, and they may also be swept backwards in the lee of some obstruction abrading to form a sharp line of separation between two smoothed surfaces. Horizontal turbulence will whirl particles of the river's load round in circles, particularly when they become trapped in a crevice or joint in the rocks of the river bed. As the pebbles swirl round and round they strike against the rock over and over again gradually 'drilling' a hole into the solid rock. Such holes are known as **potholes** (not to be confused with the caves in limestone hills referred to in *Studies E3* and *E4*) and as these enlarge and coalesce, as shown in *Fig 24*, they are seen to represent a very powerful force by which a river is able to deepen its bed.

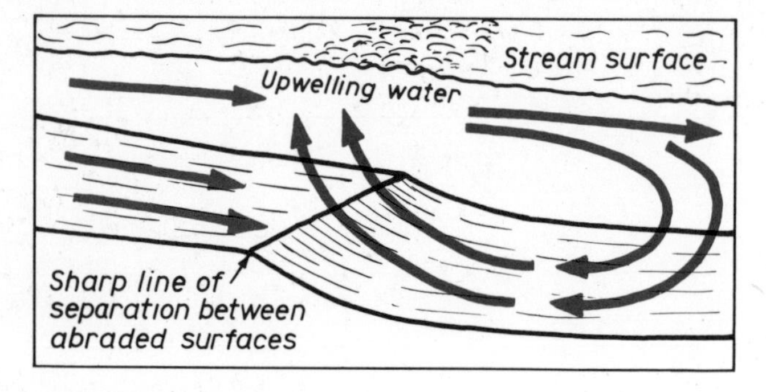

Fig 23 Vertical turbulence

Fig 24 Potholes in a stream bed

As the rock fragments of the river's load strike against the rocks of the river channel and against each other, they are themselves broken up into progressively smaller particles. Even the largest boulder may be reduced to fine gravel, not by a single blow but by the repeated small impacts each of which slightly weakens the rock structure until finally the boulder cracks. The gradual reduction in the size of the load particles achieved in this way is **attrition.**

Hydraulic action, corrasion and attrition are supplemented by **corrosion** when the rocks involved have a composition which makes them liable to solution. The processes involved are similar to those described under 'chemical weathering' in *Study A1* and are most noticeable in the case of limestone rocks.

In these ways a river deepens its bed, i.e. it carries out **vertical erosion.** If vertical erosion was the only process involved, every river would flow at the base of a gorge with vertical sides and of a width equal to that of the river. Such a situation does occur either where the rock is very resistant to weathering or where the effects of weathering are much reduced by lack of atmospheric moisture as in arid regions. Gorges also occur in places where vertical erosion has been so rapid that weathering cannot keep pace with it (*Fig 25a*). But normally, weathering, combined with mass movement of weathered material, continually wears back the valley sides and carries the weathered material to the river to be transported seaward. Since the highest level of a valley has been subjected to the weathering process for a longer period than the lower slopes, the former is worn back further, thus producing a V-shaped cross-section (*Fig 25b*), an example of which is considered in the next Study.

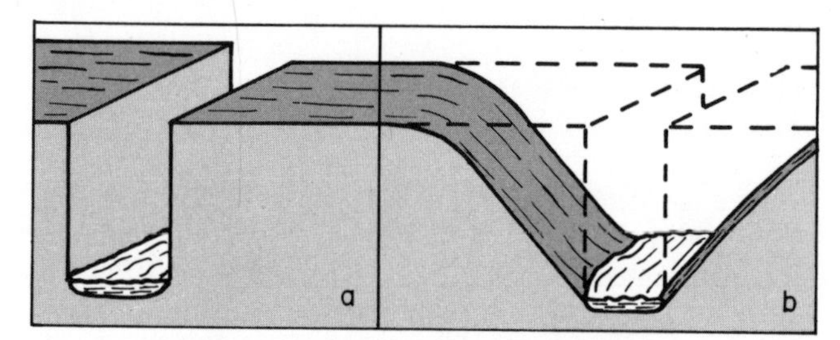

Fig 25 (a) Gorge carved entirely by river erosion
(b) V-shaped valley: volume removed by weathering and mass movement is shown by pecked lines

The increase of volume and velocity in a river in flood may greatly increase its erosive and transporting capacity, as is witnessed by such tragic events as the Lynmouth disaster in 1952 and the Whangaehu, New Zealand, floods of 1953. Changes produced under flood conditions will not normally be modified to any marked extent by long periods when the river flows at its normal volume. However it is not necessary to invoke abnormal conditions to show that rivers are capable of carving valleys. If a river was to deepen its bed by the processes which have been described at the rate of 0.5 mm per

Fig 26 Bed of the River Taff, Glamorgan

year it would erode a valley 500 m deep in one million years—a very short period on the Geological Time Scale (*Fig 1*).

tions fixed to the side of the channel produce some of the effects of turbulence.

FURTHER STUDY

Practical

(1) Observation of a dried-up stream bed, making photographs and/or sketches, may only rarely be possible but is very rewarding. Failing this, study any available photographs.

(2) The observations of stream flow suggested in *Study A3* are very relevant in this context and may be usefully carried further with particular attention to evidence of turbulence.

(3) An interesting model stream channel can be constructed using a length of plastic guttering. Gravel, sand and stones can be used to simulate the load, and obstruc-

Reading

Gresswell: *Rivers and Valleys*, Ch. 4, pp. 39–46
——: *Physical Geography*, Ch. 16, pp. 152–155
Hardy & Monkhouse: *Physical Landscape in Pictures*, pp. 39–46
Holmes: *Principles of Physical Geology*, Ch. XVIII, pp. 504–514
Horrocks: *Physical Geography and Climatology*, Ch. 5, pp. 57–58
Monkhouse: *Principles of Physical Geography*. Ch. 6, pp. 109–113
*Sparks: *Geomorphology*, Ch. 5, pp. 81–91
Strahler: *Physical Geography*, Ch. 24, pp. 347–348

Ardingly Brook, Sussex

Study A6

THE UPPER COURSE OF A RIVER

O.S. 1:25 000 (4 cm to 1 km *or* 2½ in to 1 mile)
Sheet TQ/33

At G.R. 338351 on the accompanying map extract will be found the source of a small stream (Threepoint Gill) which flows S through the High Weald of Sussex. By the southern limit of the extract it is known as Ardingly Brook and it later flows into the River Ouse to reach the English Channel at Newhaven.

MAP ANALYSIS EXERCISES

(1) Determine by a close study of the contours the altitude of the source of this stream and the altitude of the Ardingly Brook at the S edge of the map extract (G.R. 332300).

(2) Measure as accurately as possible the length of the stream between the source and the edge of the map. This can be done with an opisometer (map measurer), with dividers, by twisting the edge of a piece of paper along the stream, or with a piece of cotton. Try two methods and compare your results.

(3) You can now calculate the **average gradient** of this part of the stream by using the formula

$$\textbf{average gradient} = \frac{\textbf{horizontal distance}}{\textbf{difference in altitude}}$$

remembering that both measurements must be in the same units.

(4) The gradient calculated above is the average for this part of the stream and further study shows that the gradient is not constant. To check this, work out the gradients: (a) from the source to the footbridge at G.R. 333340; (b) from 333340 to 'Horse Bridge' (G.R. 332317) and (c) from 332317 to 332300.

(5) A more detailed picture of the slope of the river can be obtained by drawing a **longitudinal profile** (or long profile). This is prepared in a similar way to a contour section, but plotting each contour crossed by the river at the correct distance apart measured along the river.[1]

(6) Draw accurate cross-sections (**transverse profiles**) across the valley (a) from 328342 to 339336, (b) from 327307 to 341307.

(7) Study the actual shape of the stream, noticing the occurrence of small and large bends.

The preceding type of map study provides the essential facts on which to base a description of any river valley. If it is possible to visit the valley the map evidence can be checked against reality, and a description can then include minor features such as small waterfalls, local evidence of erosion in the form of potholes in the stream bed and undercut banks, any small depositional features, etc. Also field sketches and/or photographs can supplement the visual evidence of longitudinal and transverse sections.

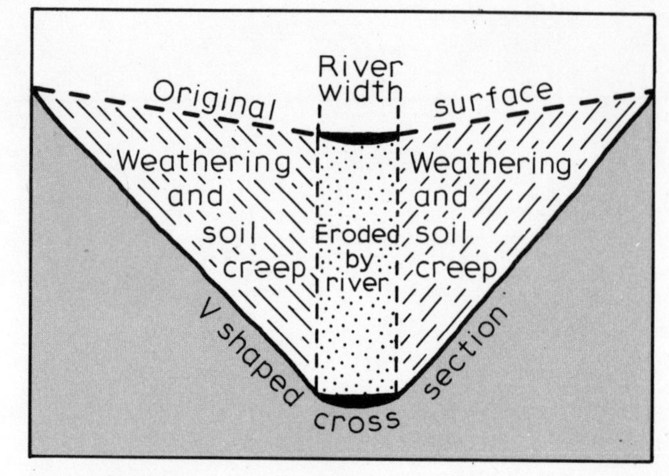

Fig 27 River erosion, weathering and soil creep combine to produce a V-shaped cross-section

The overall gradient of this portion of Ardingly Brook is of the order of 1 in 58, with the steepest gradients near to the source and gradually decreasing gradients downstream. As suggested in *Study A3* above it is dangerous to infer from this that the river is flowing faster in its upper section. However it can be said that with a steep gradient, vertical erosion is likely to be dominant and in at least the first mile of its course this river will be actively deepening its bed by those processes described in the previous Study. Weathering of the valley sides and soil creep add to the 'tools of erosion' and to the load of rock debris carried along by the river and the total effect is the erosion of a valley with steeply sloping sides and hardly any flat ground beside the river (*Fig 27*). This is referred to as a V-shaped valley, and is exemplified in the first transverse profile drawn (*Ex. 6a above*).

Following the river southward from Horse Bridge (332317) certain changes become apparent. The gradient is now markedly less, in fact the river crosses only one contour between Horse Bridge and the edge of the map. The course of the river shows curves of a larger scale than minor bends of the upper valley and the cross-section (*Ex. 6b above*) indicates that although the valley sides remain steep there is now a narrow, but significant strip of flat ground on either side of the stream. These features all show the transition from the upper to the middle course of the river. Although vertical erosion still occurs, **lateral erosion** now makes a significant contribution to the valley form, and will be considered in detail in the following Study.

[1] In order that sections of different areas drawn at different times can be easily compared it is useful for the student to adopt a consistent scale. For a section from the 1:63 360 map a vertical scale of $\frac{1}{10}$ in to 100 ft is suitable, giving a vertical exaggeration of ×5.28; and for the 1:25 000 map a vertical scale of $\frac{1}{10}$ in to 50 ft is recommended, giving a vertical exaggeration of ×4.18.

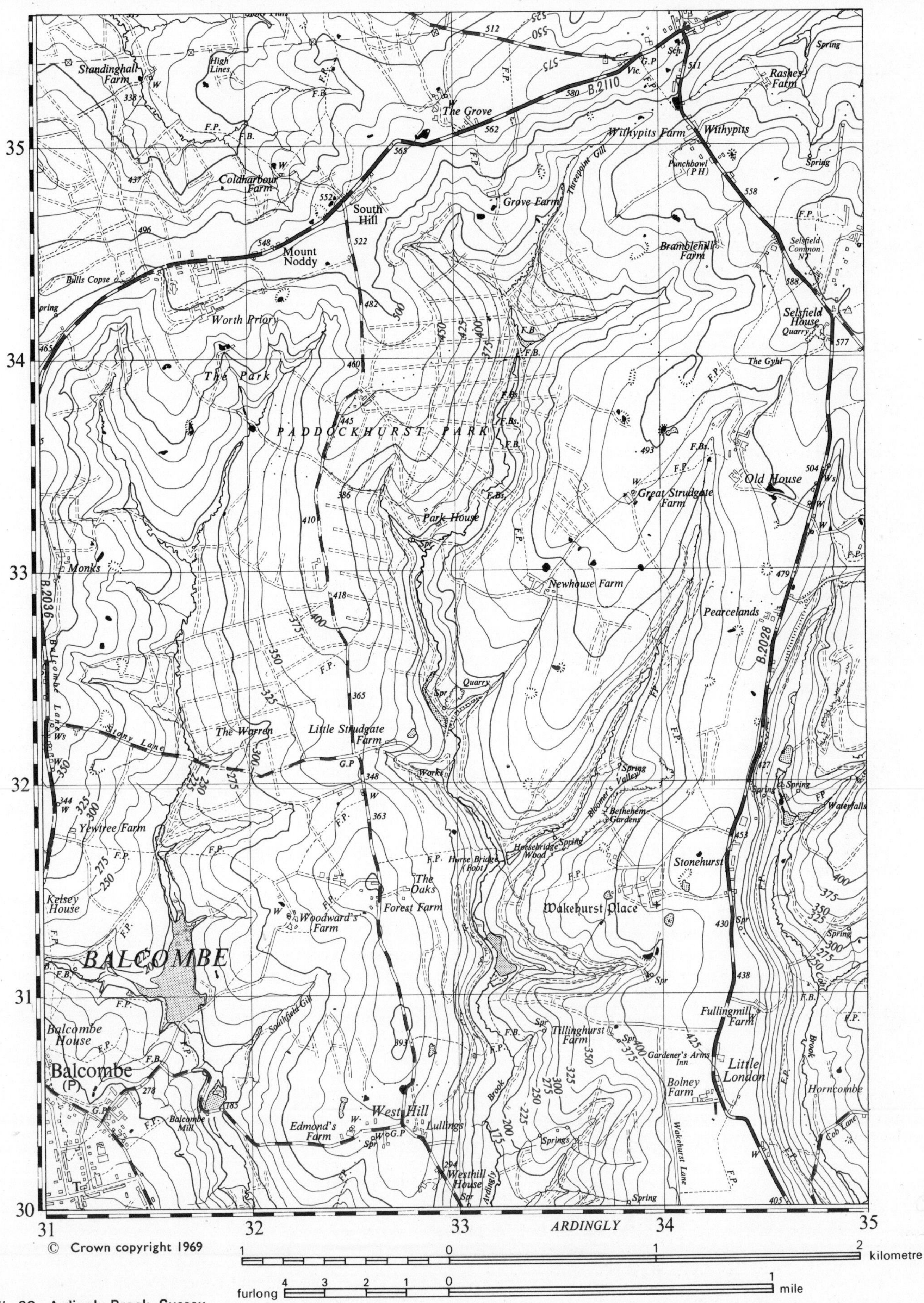

© Crown copyright 1969

Fig 28 Ardingly Brook, Sussex

Scale 1 : 25 000 (4 cm to 1 km *or* $2\frac{1}{2}$ in to 1 mile)

Contour interval 25 ft (7.5 m)

FURTHER WORK

Practical

(1) Similar exercises to those detailed above may usefully be worked for any relatively steep and narrow river valley; particularly useful examples may be found in the streams which flow to the North Devon coast, and the upper courses of rivers on Sheet 154, 1:63 360 (Cardiff) also provide good examples.

(2) If a valley of this type is locally available it should be visited and sketch maps and field sketches produced.

(3) It is not to be expected that all, or even most, rivers so studied will yield similar results to those of Ardingly Brook. Is it possible to explain any discrepancies by reference to the geology of the area concerned? Do other rivers in the same area have similar characteristics?

Reading

Carter: *Land Forms and Life*, Section 5, pp. 45–50
Gresswell: *Rivers and Valleys*, Ch. 6, pp. 59–63
——: *Physical Geography*, Ch. 16, pp. 157–159
Hardy & Monkhouse: *Physical Landscape in Pictures*, pp. 39–41
Horrocks: *Physical Geography and Climatology*, Ch. 5, pp. 59–64
Monkhouse: *Principles of Physical Geography*, Ch. 6, pp. 120–123
——: *Landscape from the Air*, pp. 18, 19
Scovel: *Atlas of Landforms*, p. 20
*Sparks: *Geomorphology*, Ch. 5, pp. 91–95
Strahler: *Physical Geography*, Ch. 24, pp. 351–352; Ex. 1, p. 364
Wooldridge & Morgan: *An Outline of Geomorphology*, Ch. XII, pp. 140–142

The Ribble Valley above Preston, Lancashire

Study A7

A RIVER IN ITS MIDDLE COURSE

O.S. 1:63 360 (1.6 cm to 1 km *or* 1 in to 1 mile)
Sheet 94, Preston

The R. Ribble rises on the slopes of Whernside in the Central Pennines and follows a southerly course through Ribblesdale (*Study E4*) before turning SW to enter the Irish Sea at Preston.

MAP ANALYSIS EXERCISES

(1) The 100 ft contour (30 m) crosses the R. Ribble at 698377; the 50 ft contour (15 m) crosses at 621338. Measure the distance along the river between these two points and work out the average gradient. Compare your result with that for Ardingly Brook (*Study A6*).

(2) Draw a map of the portion of the R. Ribble shown in *Fig 30*. Lightly shade the area of apparently flat land[1] bordering the river and indicate by hachure symbols those places where the river is bordered by steeply rising land.

(3) Measure the width of the **meander belt** (a) in the eastern portion of the map, (b) in the western portion. (The meander belt is the width between the extremes of the river curves; *Fig 29*.) Similarly measure the width of the flat valley floor—**the flood plain**—in the same parts of the area. How do your answers compare with similar measurements at the S edge of the Ardingly Brook map? (*Note*. Maps differ in scale.)

[1] The fact that the 50 ft contour (15 m) runs closely parallel to the R. Ribble for a considerable distance does not mean that the river is set well below the flood plain. Remember that all the contour signifies is that the river is less than 50 ft above sea level (perhaps 49 ft) and the valley floor is above 50 ft (perhaps 51 ft).

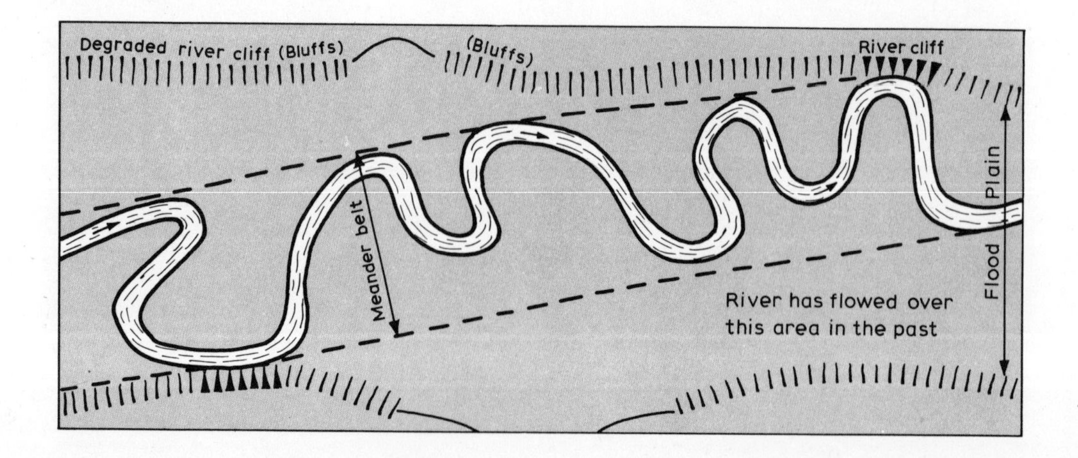

Fig 29 Meandering river and its flood plain

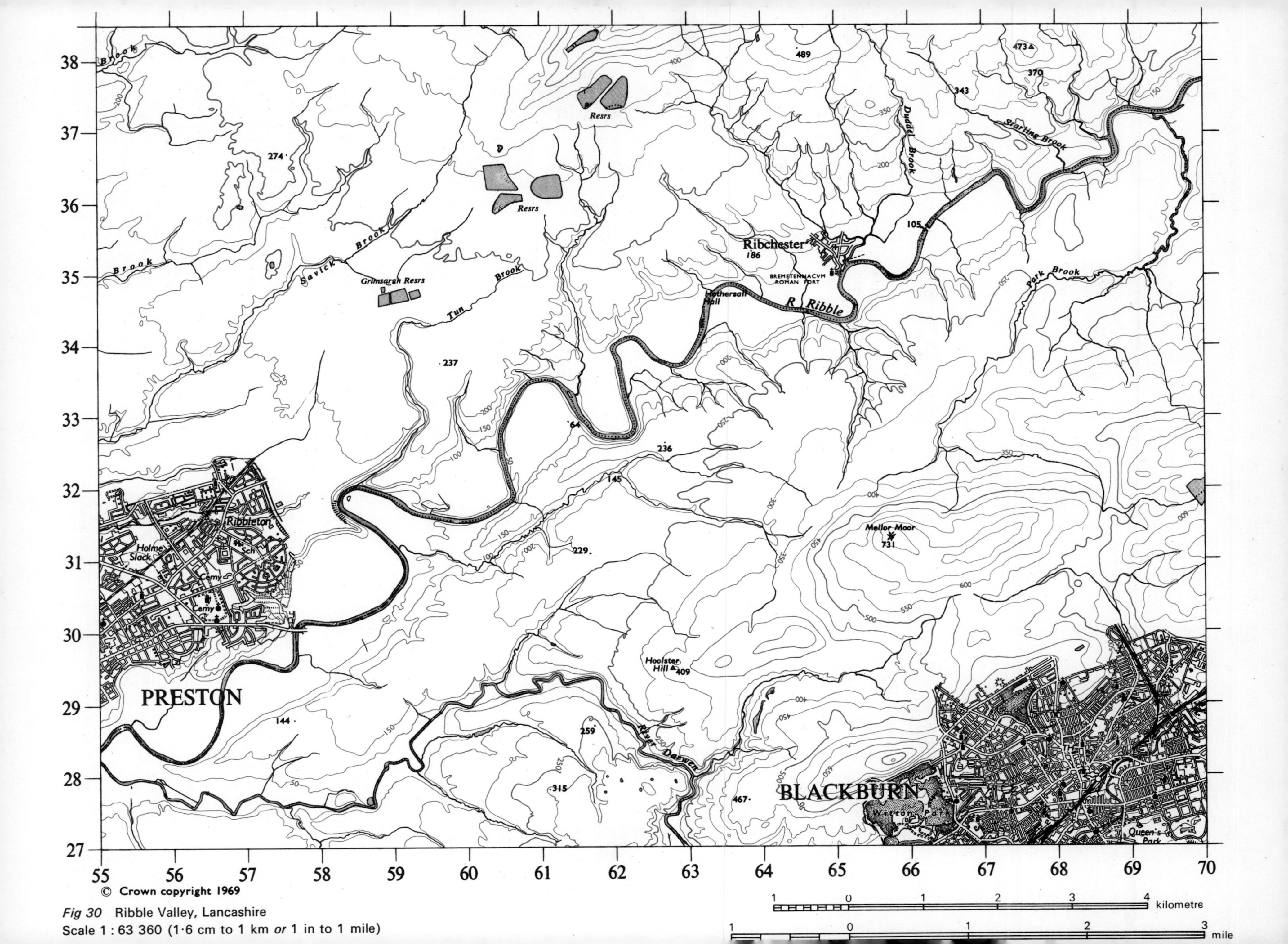

Fig 30 Ribble Valley, Lancashire
Scale 1 : 63 360 (1·6 cm to 1 km *or* 1 in to 1 mile)

(4) Refer to the map drawn for question (2) above: Study carefully those places where you have indicated steep slopes rising directly from the river banks; then try to make some general statement which describes the position of such steep slopes (or **river cliffs**) in relation to meanders.

In this section of the Ribble Valley the meander form of the river is clearly a major feature of the land forms. When the flow of water over a sand pile was studied (*Study A3*), it was noted that curves or meanders formed. On occasions it may be possible to relate the initiation of such curves to some minor obstruction. Such obstructions are not essential to the formation of meanders however, for any liquid, or gas, flowing in contact with a surface of different density is liable to oscillate. This may be simply demonstrated by observing a trickle of water across a sloping sheet of smooth glass. The same principle leads to the breaking up of a cylindrical column of water from a tap and is also involved in the turbulent motion of wind over water which generates waves.

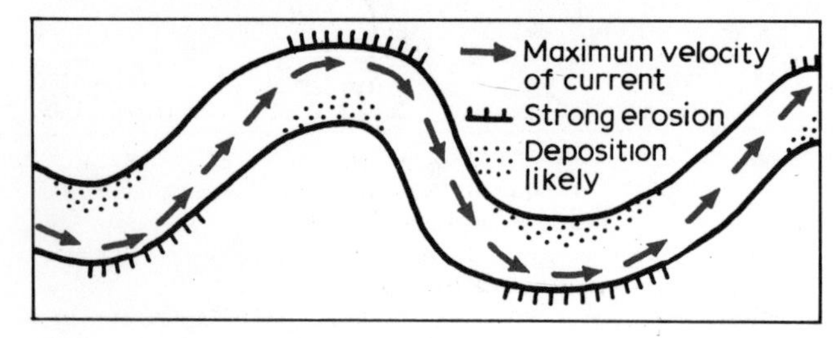

Fig 31 Water flow round meanders

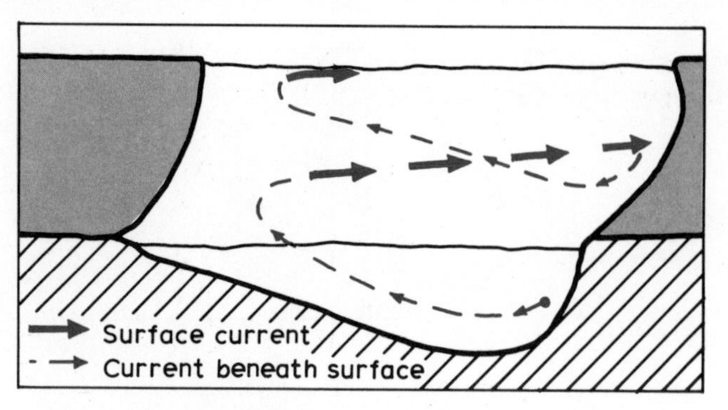

Fig 33 'Corkscrew' flow in a meander

Once established, a stream meander becomes self-perpetuating. Water in any stretch of the river flows in a straight line until it strikes a bank and is diverted (*Fig 31*). Thus the main surface current is always directed towards the outside, and the downstream side, of a curve, while the motion of water on the inside of curves is much less powerful. If the movement of water at depth is considered also, the river will be seen to progress downstream in a corkscrew like motion (*Fig 33*). Since active erosion is to be expected where the movement of water is most powerful, it can readily be seen that the outside and downstream sides of meanders will be worn away while the load so produced is likely to be deposited on the inside bank of a curve. Similarly the cross-section of the river channel will normally be asymmetrical in form with a steep **undercut bank** on the outside, and a gently sloping depositional **slip off** slope on the inside of each meander. *Fig. 32*

The total effect of the above features is considerable, and may be seen if *Fig 34* is studied. (Better still, draw

Fig 32 Undercut banks and slip off slope, Mimms Hall Brook, Hertfordshire

Fig 34 Formation of flood plain by migration of river curves

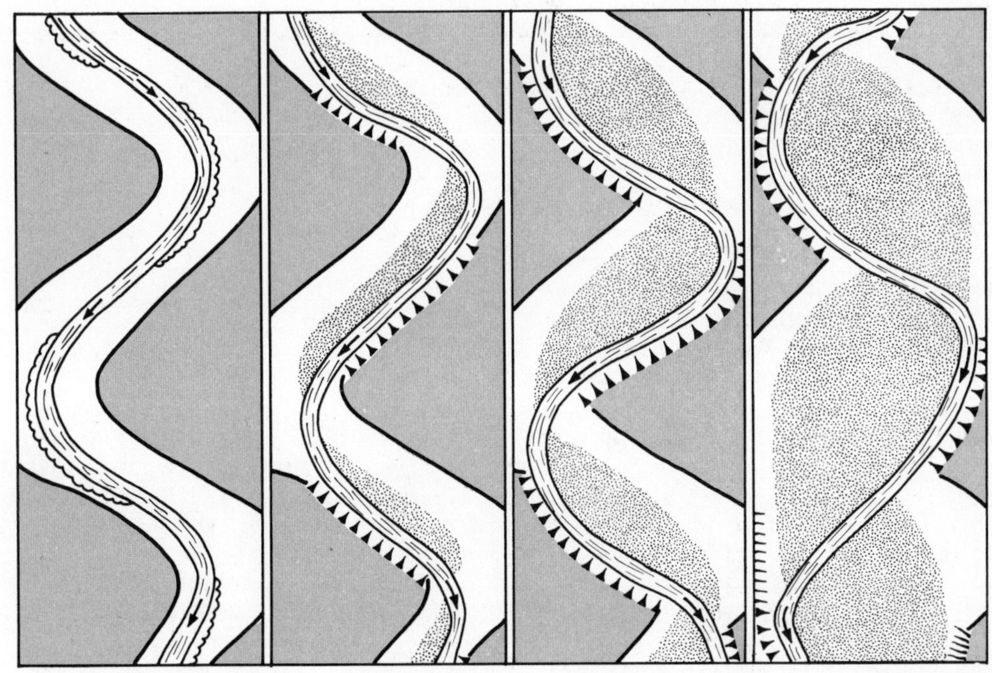

Stage I. River curving between interlocking spurs. Maximum erosion on outside of curve marked

Stage II. River beginning to erode spurs and form river cliffs ▼▼▼▼. Deposition on inside curves

Stage III. Curves more pronounced. Downstream migration of curves has reduced spurs to cuspate form

Stage IV. Spurs largely removed. Extensive flood plain formed. Old river cliffs degraded. Deposition in areas crossed by river at some stage

the diagram for yourself stage by stage noting the changes at each stage and the reasons for them.) The increased amplitude and downstream migration of curves gradually reduces the spurs first to a **cuspate** form and eventually removes them altogether so producing a broad flood plain across which the river can swing without obstruction.

(5) The gradual reduction of spurs may be illustrated from the Ribble Valley. Draw annotated contour maps of spurs and river curves: (a) area around 675360 to show a spur as yet little affected by erosion; (b) area around 620327 to show cuspate spur jutting towards Balderstone Hall with a steep **river cliff** being actively eroded on its eastern side; (c) area around 585315 to show active erosion of river cliff and last remnant of **cuspate spur.**

The application to the Ribble Valley of the concept of meander development suggested in this section, can be extended to produce an impression of future changes, and it can be seen that when the remaining spurs have been removed the Ribble will meander freely across a broad trench flanked by slopes which will be in part at least the weathered remains of river cliffs. Such a valley form is considered in detail in the following Study.

FURTHER STUDY

Practical

(1) Study of maps will reveal rivers showing downstream migration of meanders and partial removal of spurs—e.g. R. Avon, N of Salisbury (O.S. Sheet 167, 1:63 360).

(2) Careful observation of most streams, even very small rivulets, will provide exemplification of the water currents produced in river curves. These should be sketched and the dimensions of the curves (wave length, amplitude, stream width, etc.) measured. If a number of streams are so studied it may be possible to establish a fairly constant ratio between river width and meander 'wave length'. Some research workers suggest that this ratio is approximately 1:13, though of course local variations may occur, especially along a small stream.

Reading

Carter: *Land Forms and Life*, Section 6, pp. 51–54

Dury: *The Face of the Earth*, Ch. 8, pp. 88–98

Gresswell: *Rivers and Valleys*, Ch. 6, pp. 64–67

——: *Physical Geography*, Ch. 16, pp. 159–161, 166–169

Hardy & Monkhouse: *Physical Landscape in Pictures*, p. 47

Holmes: *Principles of Physical Geology*, Ch. XVIII, pp. 528–534

Horrocks: *Physical Geography and Climatology*, Ch. 5, pp. 64–66

Monkhouse: *Principles of Physical Geography*, Ch. 6, pp. 123–124

Scovel: *Atlas of Landforms*, pp. 86–87

*Sparks: *Geomorphology*, Ch. 5, pp. 95–99

Strahler: *Physical Geography*, Ch. 24, pp. 352–354; Ex. 2. p. 364

Wooldridge & Morgan: *An Outline of Geomorphology*, Ch. XII, pp. 142–143

A RIVER IN ITS LOWER COURSE

O.S. 1:63 360 (1.6 cm to 1 km *or* 1 in to 1 mile)
Sheet 112 *or* 121 *or* 122,
O.S. 1:25 000 (4 cm to 1 km *or* 2½ in to 1 mile)
Sheet SK/64 and SK/63

Fig 35 **Field sketch of river cliffs, Radcliffe on Trent**

As it passes Radcliffe-on-Trent the R. Trent is flowing through a flat floored trough about 3 km in width, the limits of which are approximately defined by the 100 ft contour (30 m), though the greater part of the valley floor lies at an elevation of about 18 m above sea-level as shown by spot heights of about 60 ft. The features of the valley can be well seen from the cliffs (649401) of red Keuper Marl which give Radcliffe its name (*Figs 35, 37*).

The river meanders from one side of the valley to the other, and, as at Radcliffe, it occasionally reaches to the extreme edge of the flood plain. At such points river cliffs are found. At Radcliffe these take the form of a nearly vertical face cut in Keuper Marl which at this point dips very gently in a SE direction. On such a steep face being actively undercut by the river, the soil cover is naturally sparse, yet trees cling to the cliff forming an almost complete cover of woodland. Many of these trees show evidence in their curving trunks of active soil creep (*see p. 5*).

To the SW and NE of this steep cliff the flood plain is bounded by gentler, soil covered slopes with a more stable vegetation (*see map extract, Fig 38*).

After periods of heavy rain or following rapid snow melt it may be difficult to pick out the course of the river. All the land between the railway embankment (640396) and the river channel will be flooded and even after the water has apparently disappeared the water table in the flood plain meadows is very near the surface and the alluvial soil can be seen to show mud cracks and marshy hollows. Flood water does not normally extend on to the left bank of the river at this point due to the construction of flood preventive embankments.

Close observation of the land between the railway and the river reveals a slight, but significant, slope downwards away from the river. This is a miniature form of

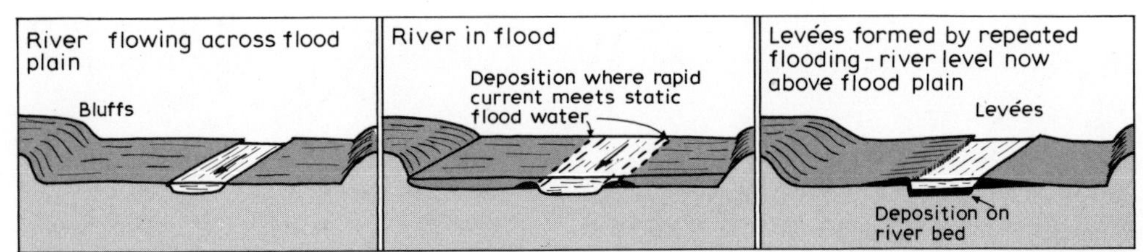

Fig 36 Formation of levees

levée formation which becomes a very important feature in the land forms of many of the world's great rivers. The small stream which joins the Trent at the Ferry (645398) first enters the flood plain south of Holme Pierrepont (623380) where it is heading N as if to join the Trent near Colwick. Instead it turns E for some 3 km before its confluence. Other streams can be seen to follow unusual courses; for example, one commences within 180 m of the R. Trent at 652413 but flows NE for 3 km through Shelford before joining up at 663430.

Apart from levées and artificial flood banks the flood plain seems almost completely flat. However it should be noticed that the villages of Stoke Bardolph and Shelford are sited on land which lies some 2 m above the general level of the plain and are thus protected somewhat from flooding.

The Trent Valley at Radcliffe represents a stage in valley formation following those described in the previous Studies. Here the meander belt approximates in width to the flood plain so that no spurs of high ground jut out into the flood plain to obstruct the free downstream migration of meanders; a form which may be taken as marking the transition from a mature to an old-age valley (*Study A10*). The existence of the river cliff which is being actively undercut by the river indicates clearly that the Trent, at this point, has not yet removed all obstacles to its free development, and comparison may be made with other rivers which have formed a flood plain many times the width of the meander belt. In these cases the whole meander belt may very slowly migrate from side to side across the flood plain.

As the Trent meanders migrate slowly downstream they leave behind vertical river cliffs which become **degraded**, or rendered less steep by weathering and soil creep.

Few geomorphological features are quite as simple as they may at first appear, and this is particularly true of the British Isles where the climatic and landscape changes and variations in sea-level during the Quaternary Ice Age have influenced a wide range of landscape features. At any level of study, therefore, there are features which require further investigation, and ideas of possible modes of formation must be carefully related to observation and measurement. The Trent Valley at Radcliffe provides a good example of the complexity of landscape features. Research work at an advanced level indicates that this trough was cut by the R. Trent at a time in the last phases of the ice age when it was diverted from a more southerly course by extensive ice sheets. The river terraces upon which Stoke Bardolph and Shelford are sited are related to changes in the nature of the river at a similar period. But even if such advanced studies are taken into account, it is still valid to consider the Trent Valley as the natural consequence of the further continuation of the processes of spur removal described on the R. Ribble in the previous Study.

At Radcliffe the Trent has an average gradient of 1 in 2450 (about 0.6 m/km) and a normal surface velocity of about 2.5 km/h. It is subject to frequent floods, even though these have now been reduced by river control works. At such times a shallow sheet of almost stagnant water extends across the flood plain and the river itself carries along a heavy load of rock debris. Where the fast-moving river water meets the stagnant flood water a series of eddy currents are formed and some of the river's load is deposited (*Fig 36*). This deposition of silt occurs to some degree all over the flood plain, but most noticeably along the river banks to form the **levées**. In the case of some of the world's great rivers this process combined with deposition on the river bed has served to gradually raise the whole river above the flood plain.

Fig 37 Trent Valley at Radcliffe; from GR 647399 looking WNW

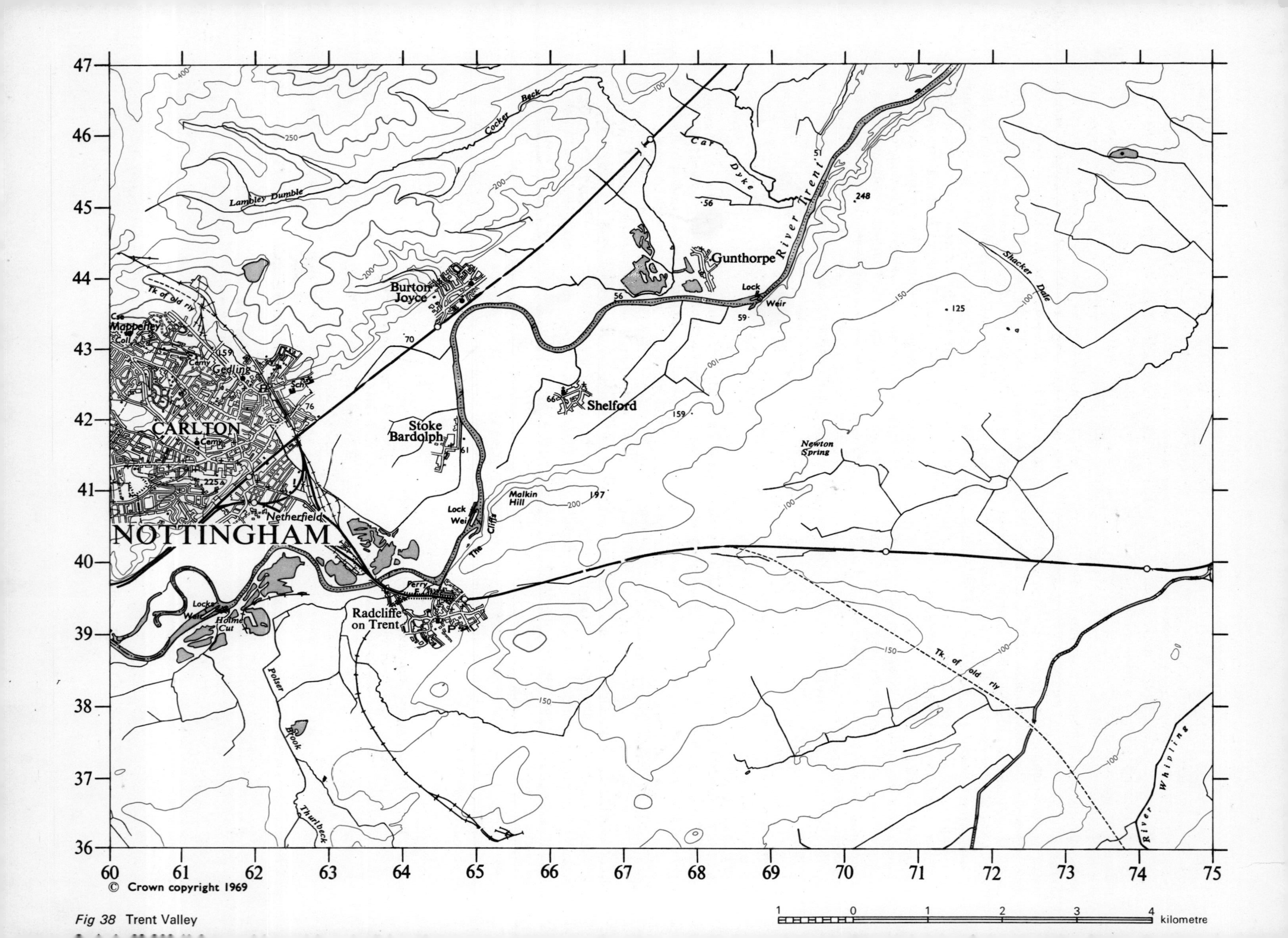

Fig 38 Trent Valley

The formation of levées, even if only of small magnitude as in the case of the Trent, inevitably leads to a number of other features on the flood plain because water cannot flow up a slope and free drainage to the main river is no longer possible. Flood water remains for a long time before percolating downwards and leaving marshy areas. More significantly it is no longer possible for small tributaries to join the river at the obvious point. Instead such a tributary as the Polser Brook flows parallel to the river before joining at a point further downstream, by which time the river has fallen sufficiently to permit the confluence. Similarly the presence of levées accounts for surface water collecting near the river and then flowing in a brook away from the river, to join it at a point further downstream.

It has already been shown that lateral erosion causing the downstream migration of meanders as described in *Study A7* plays a vital part in widening the river valley and removing spurs and other obstructions to the free sweep of the river across its flood plain. Lateral erosion is also responsible for minor features of the flood plain. It may be noted in passing that the distinction between a steep undercut bank and a gentle slip off slope is not apparent in this section of the Trent, nor on other rivers which are regularly used for navigation where the wash from vessels causes undercutting and steep banks on both the inside and outside of a curve. Flood prevention work in general is aimed at stabilizing the river's course, but evidence can be found in the form of marshy hollows and patches of standing water that the river's course has not always been the same. Meanders change position and shape, and **ox-bow lakes** and **cut offs** are formed (*see Fig 37*). Sheets 112 and 121 show a good example of such a feature to the west of Nottingham, at Sawley (465315) where it is interesting to note that the county boundary follows the abandoned loop of the ox-bow rather than the present river channel. Unfortunately this interesting feature has been obliterated by M1 Motorway works.

With the continual changes of meander form and position under natural conditions it becomes clear that at some past time the river has flowed over each part of the flood plain, often many times, depositing **alluvium** to build up considerable thickness of such material.

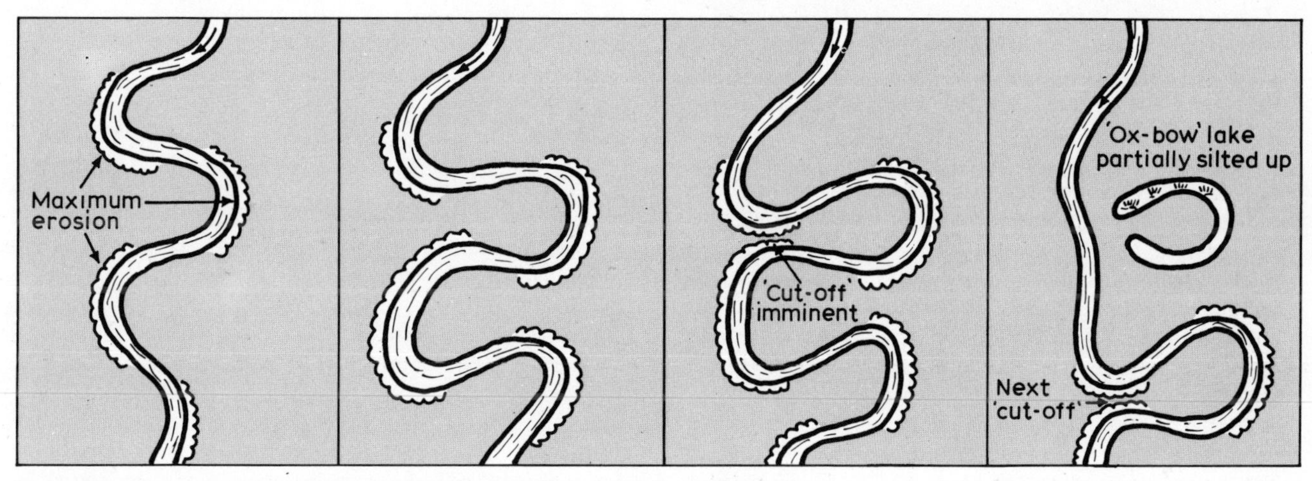

Fig 38a Stages in meander development showing formation of 'Ox-Bow' lake

FURTHER STUDY

Practical

(1) Draw an annotated sketch map of the section of the R. Trent shown on the map extract (*Fig 38*). All the features mentioned in the text should be carefully identified and labelled.

(2) If a large river with a wide flood plain is accessible, field study should be made to seek resemblances to, and variations from, the features of the R. Trent.

(3) Similar exercises can be worked from other large-scale maps showing rivers which have a well-developed flood plain. If large-scale maps can be obtained of parts of the Mississippi or other rivers larger than those of Britain, the characteristic features described in this Study can be seen on a larger scale.

Reading

Carter: *Land Forms and Life*, Section 7, pp. 55–61
Dury: *The Face of the Earth*, Ch. 8, pp. 88–98
Gresswell: *Rivers and Valleys*, Ch. 6, pp. 68–69
——: *Physical Geography*, Ch. 16, pp. 166–169
Hardy & Monkhouse: *Physical Landscape in Pictures*, p. 48
Holmes: *Principles of Physical Geology*, Ch. XVIII, pp. 535–545
Horrocks: *Physical Geography and Climatology*, Ch. 5, pp. 66–69
Monkhouse: *Principles of Physical Geography*, Ch. 6, pp. 124–127
——: *Landscape from the Air*, p. 21, p. 22
Scovel: *Atlas of Landforms*, pp. 16, 88–89, 97
*Sparks: *Geomorphology*, Ch. 5, pp. 95–99
Strahler: *Physical Geography*, Ch. 24, pp. 355–357; Ex. 3, p. 366
Trueman: *Geology and Scenery*, Ch. 5, pp. 71–74
Wooldridge & Morgan: *An Outline of Geomorphology*, Ch. XII, pp. 156–158

The Mouth of the River Rhône, France

Study A9

DELTAIC FORMATIONS

Carte de France 1: 250 000 (4 mm to 1 km *or* ¼ in to 1 mile) Sheet NK 31–3 Marseille

If the rivers studied in the three preceding Studies are traced to the sea, they can be seen to widen gradually and merge with the sea in an **estuary.** This is the typical feature of river mouths of the British Isles and of many large areas of the world, and may easily be related to the 'drowning' of the lower portions of river valleys consequent upon a rise of sea-level following the Pleistocene Ice Age (*see Study B8*). These estuaries are kept largely clear of river sediment by the strength of the scouring action of the tides. Where tidal scour is less effective or where other factors are dominant **a delta** is a more common type of river mouth.

The map (*Fig 39*) shows the delta of the R. Rhône. This river divides into two main branches (or **distributaries**) 4 km N of Arles. The E branch can be identified as the larger, both by its width on the map and by its name (Grand Rhône). In fact it carries some 85% of the R. Rhône's water to the Mediterranean. The river at Arles is 2 metres above sea-level showing the very slight gradient between here and the sea—a distance of 47 km. (Calculate the average gradient).

The representation of small islands and probable abandoned channels suggests the occurrence of deposition along the Grand Rhône. On the original map from which *Fig 39* was prepared, certain human features are shown which have a significance in a study of the physique of this area. The majority of settlements, roads and railways in the delta area are normally located relatively near to the river, whereas land further back from the river is marshy or covered by lakes. This fact, together with the brown, cross-hatched lines alongside the rivers suggests the presence of levées—further evidence of deposition. The W branch (Petit Rhône) exhibits similar features, with a rather more meandering course. A sharp change of direction 12 km from the sea suggests that the river has changed course in the past, and the Canal de Pecais Sylvéréal may be used to reconstruct the earlier line. Another change of course, this time for the Grand Rhône, is suggested by the inlet of the Vieux Rhône 18 km W of the present mouth. Thus it seems likely that the river outlets are steadily migrating eastward.

Between the two branches of the Rhône lies the Camargue, a flat region with many marshes and lakes (étangs). The largest of these, the Etang de Vaccarès, is no more than 1 m deep, and like the other étangs receives the greater part of its water by the accumulation of rain water into the basin formed by the slightly raised river banks. The Etangs are separated from the sea by sand bars and dunes. What evidence of coastwise drift of sand can be observed? In which direction(s) is beach material drifting?

A million years ago, at the close of the Tertiary period, the River Rhône flowed into a triangular embayment of the sea extending SW towards Montpellier and SE to Fos. The majority of the land shown on the map (*Fig 39*) has been formed since that time mainly as the result of the deposition of the very heavy load of the Rhône. This, at present, amounts to 17 million m^3 per annum—or 50 tonnes per minute! The Mediterranean is a virtually non-tidal sea, and there are thus no twice daily currents to flush away this mass of the residues of the erosion of a large area of Switzerland and Southern France. In addition the Mediterranean waters are highly saline, and a physico-chemical reaction between this salty water and the mud-laden river water coagulates the mud particles to form larger aggregates which are too heavy to remain in suspension. Rapid accumulation of alluvium is caused by a combination of (a) the sudden checking of fairly rapidly flowing water on entering the sea, (b) coagulation of fine clays on contact with saline water, and (c) the absence of tides to sweep debris further out to sea.

The first stage in the formation of such a delta is the deposition of sandbanks in the original estuary-type river mouth; a process of which the beginnings may be seen in many larger British estuaries. As such sand banks collect they cause the river channel to divide. These channels are then similarly divided until the river is broken up into a large number of distributaries before reaching the sea. Each distributary builds up its levées, in the same way as those described in the previous Study. This accounts for the slightly higher land near to the Grand Rhône and Petit Rhône noted above as sites for human occupation, and also for the drainage towards marshy hollows lying between the main distributaries. Local conditions may modify this pattern. The Rhône, for instance, retains only two major outlets and other forms exist which will be outlined at the end of this Study.

The action of the sea upon the river's load is also of importance. Longshore drift (*see Study B3*) and currents shift material along the coast forming sand bars, and winds pile up this material into dunes. In the case of the Rhône these can be clearly seen to have isolated the shallow lagoons, or étangs, which then slowly fill with sediment to form marshy areas (e.g. Marais de la Grand Mare) which may later be drained by man.

Each delta has its own individual features brought about by local conditions. The movement of sediment by the sea and the significance of tidal currents has already been mentioned. To these may be added the relative density of river and sea water due to temperature and salinity. If the river water is less dense it will flow over the sea surface, moving quite freely forward but spreading out slowly sideways. At the edges of such a flow the river water is checked and the load deposited, thus building out levées into the sea and forming **birds' foot delta** of the Mississippi type (*Fig 40*). An **arcuate delta** of the Rhône or Nile type (*Fig 41*) is likely to form when

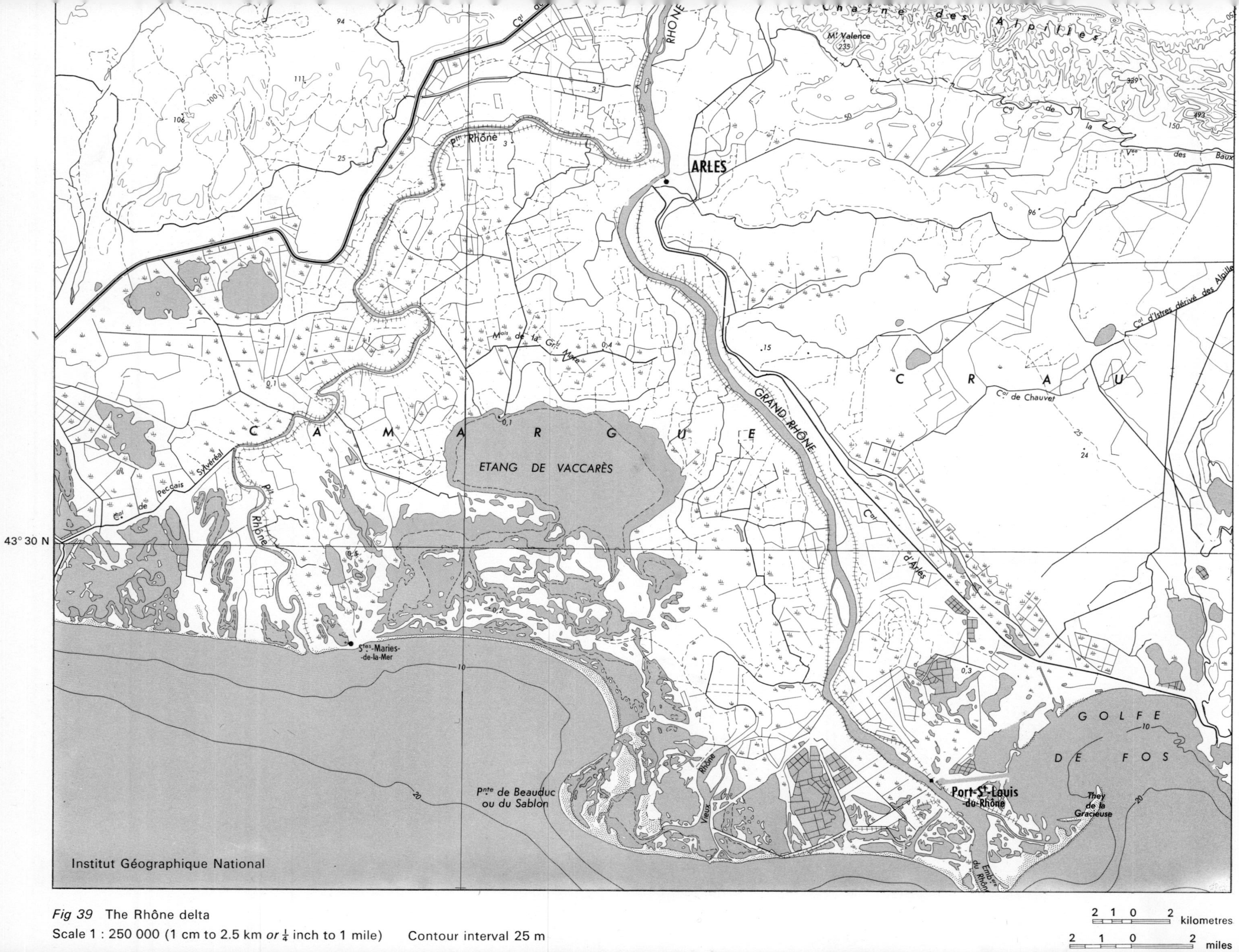

Chaîne des Alpilles
Mt Valence
RHÔNE
ARLES
GRAND RHÔNE
Cal de la Vée des Baux
Cal d'Istres dérivé des Alpilles
C R A U
Cal de Chauvet
Pit Rhône
Mais de la Gde Mare
C A M A R G U E
ETANG DE VACCARÈS
Cal de Peccais
Sylvéréal
Cal d'Arles
43° 30 N
Stes-Maries-de-la-Mer
Pnte de Beauduc ou du Sablon
Vieux Rhône
Port-St-Louis-du-Rhône
Embure du Rhône
GOLFE DE FOS
They de la Gracieuse
Institut Géographique National

Fig 39 The Rhône delta
Scale 1 : 250 000 (1 cm to 2.5 km *or* ¼ inch to 1 mile)   Contour interval 25 m

2 1 0 2 kilometres
2 1 0 2 miles

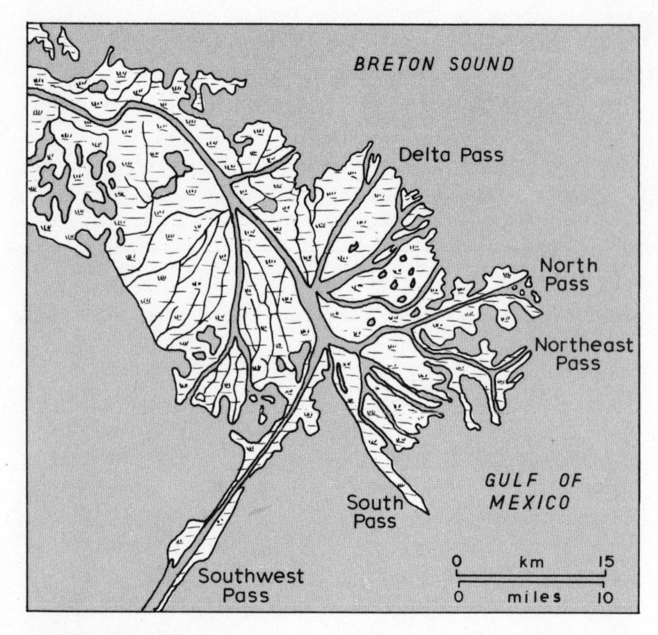

Fig 40 The Mississippi delta is of the branching, bird-foot type with long passes

river and sea waters are of similar density.

A further variant affecting the form of deltas is the degree of crustal subsidence which is occurring, partly at least in response to the enormous accumulation of sediment. When subsidence is rapid it is clear that, in general, the delta will not grow seaward with the same rapidity as if there was no submergence. The extent of the downwarping of the crust in some delta areas can be gauged from the thickness of sediments which occur—at least 3000 m in the case of the Nile.

Deltaic formations may also occur when rivers enter a lake. In such lacustrine deltas the general features of marine deltas may be observed but the absence of significant wave action, and the fact that the fresh lake water does not coagulate the finer sediment, leads to the formation of deltas of slightly differing type. Also the silt-laden river is of higher density than the clear lake water so that the inflowing current sinks to the bottom and travels along the bed of the lake as a turbidity current carrying its load with it and depositing material far out into the lake. Hence lacustrine deltas often extend far outwards from the shore at a very gentle average gradient.

Fig 41 The Nile delta has an arcuate shoreline and is triangular in plan

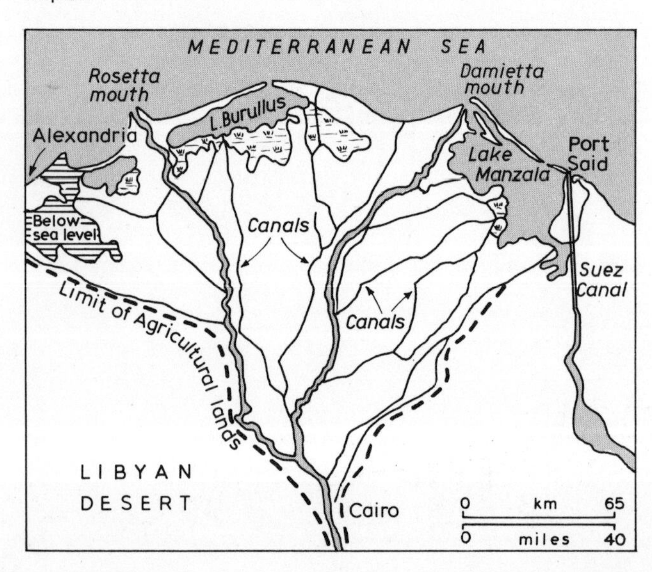

FURTHER STUDY

Practical

(1) Make clearly annotated maps of deltas based on the largest scale map available: a good atlas will contain useful maps of the Rhine–Maas, Ganges, Danube deltas, etc.

(2) Describe briefly the world distribution of deltas and account for the facts you observe.

(3) In order to appreciate the size of major deltas, it is revealing to superimpose a map of the Ganges delta on to a British Isles map of the same scale.

(4) Study a 1 : 63 360 map of the Lake District, or Snowdonia or a similar area and sketch examples of the many small lacustrine deltas which can be seen where fast flowing streams enter the lakes.

(5) If the Carte de France map of the Rhône delta is available, use spot heights to work out the slopes of the land in the triangle bordered by Lamanon (in the extreme NE), Arles and Fos. This area of La Crau represents an alluvial fan, largely of stones, deposited by the fast flowing Alpine river, the Durance, when in the past it emerged from the gap at Lamanon. (Today the Durance turns NW at this point and joins the Rhône below Avignon—*see atlas map*.) Alluvial fans have close links with deltas—both are depositional features resultant upon a sudden check in stream velocity, both frequently exhibit a river divided into many channels. The main differences lie in the fact that alluvial fans are not deposited under water and normally have a much steeper angle of slope (*see Studies D2 and D3*).

Reading

Anderson: *Splendour of Earth*, pp. 272–274, 281–283
Carter: *Land Forms and Life*, Section 8, pp. 62–68
Dury: *The Face of the Earth*, Ch. 7, p. 87; Ch. 9, pp. 108–110
Gresswell: *Rivers and Valleys*, Ch. 9, pp. 97–108
——: *Physical Geography*, Ch. 16, pp. 171–173
Hardy & Monkhouse: *Physical Landscape in Pictures*, p. 49
Holmes: *Principles of Physical Geology*, Ch. XVIII, pp. 545–554
Horrocks: *Physical Geography and Climatology*, Ch. 5, pp. 70–71
Monkhouse: *Principles of Physical Geography*, Ch. 6, pp. 131–139
——: *Landscape from the Air*, pp. 46 and 48
Scovel: *Atlas of Landforms*, pp. 98, 99, 101, 102
Strahler: *Physical Geography*, Ch. 24, pp. 358–361

THE DEVELOPMENT OF RIVER VALLEYS AND LANDSCAPES

Study A10

The preceding Studies of river valleys have suggested that a journey along a river from source to mouth will reveal a number of stages or sections:

(1) **An upper (or Torrent) Section**, where the gradient is steep, vertical erosion dominates and the river flows in a steep-sided, V-shaped valley;

(2) **A middle (Valley) Section**, in which the gradient is less steep, lateral erosion is more significant than vertical, and the floor of a less steep sided valley contains a flood plain of increasing width; and

(3) **A lower (or Plain) Section**, with only a very slight gradient towards the sea, and in which almost all erosion which occurs is balanced by deposition nearby, and where the river meanders freely across a wide flood plain in a shallow valley.

These stages are not clearly defined, but merge one into the other, and there are many local exceptions—some of which will be considered in later Studies. Nevertheless so many rivers do approximate to this pattern that one may accept that the concept emerges of an idealized long profile in which the gradient progressively decreases from source to mouth (*Fig 42*).

There is clearly a lower limit, or **base level**, to the downcutting by a river. Since water cannot normally flow uphill, this must be sea-level for the vast majority of the world's rivers. There are rare exceptions such as the R. Jordan, the base level for which is the Dead Sea, 390 m below sea-level. Like other similar basins of inland drainage, this owes its origin to crustal movement and is not primarily an erosive feature. The long profile of a river is thus tied to a base level which is normally sea-level. Any change in relative sea-level will be followed by modification in the long profile (*see Study A11*).

Within the course of a river, lakes or an outcrop of hard rock may act as a **local base level**, in which case that portion of the river above the feature will be unable to erode below this level (*Fig 43*).

Downstream from the interrupting feature the gradient of the river steepens, with a consequent increase in erosion which is partly in the form of headward erosion. The lake outlet is therefore lowered or the waterfall removed (*see Study E9*) and the irregularity in the long profile gradually disappears.

Contemporary thought in Geomorphology lays great emphasis on the **processes** by which the landscape is being formed. Many of the processes of erosion, transport and deposition by streams and rivers are still imperfectly understood and much research work is aimed at providing the sound experimental and observational data upon

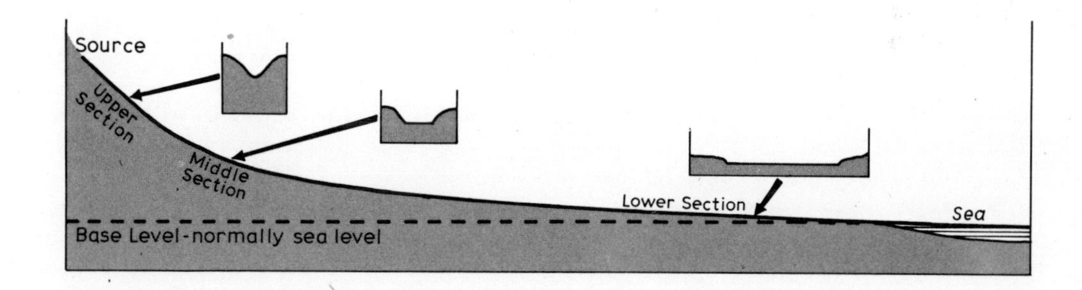

Fig 42 Long profile and cross sections of idealized river valley

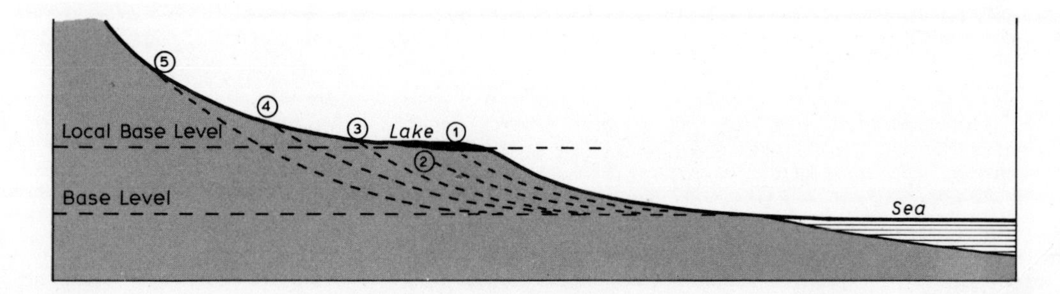

Fig 43 Local base level formed by lake and stages (1–5) in its removal

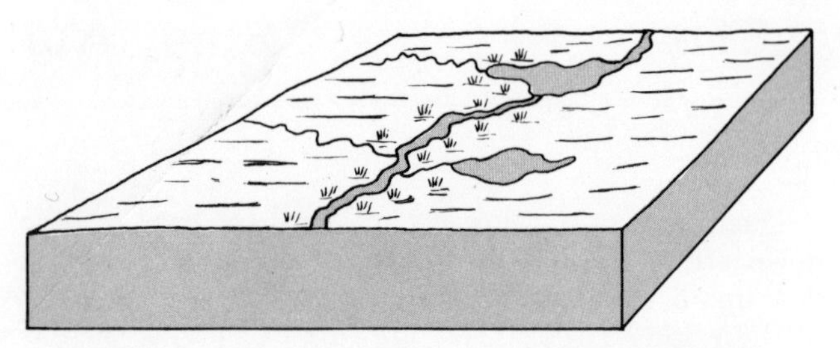

A. In the initial stage relief is slight, drainage poor

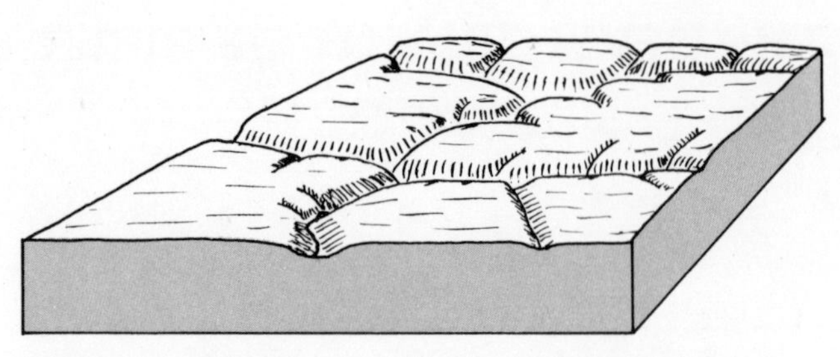

B. In early youth, stream valleys are narrow, uplands broad and flat.

C. In late youth, valley slopes predominate but some interstream uplands remain.

D. In maturity, the region consists of valley slopes and narrow divides.

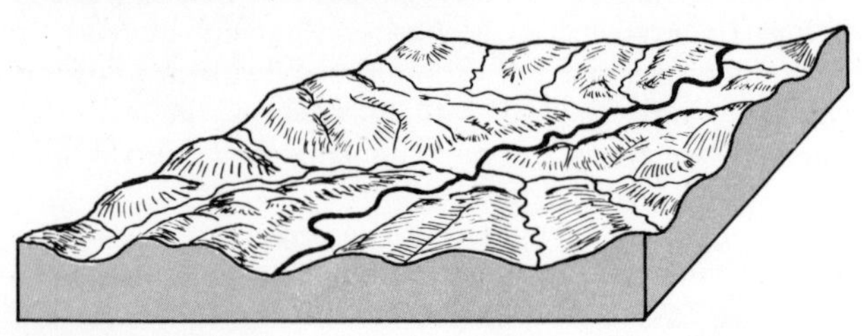

E. In late maturity, relief is subdued. valley floors broad.

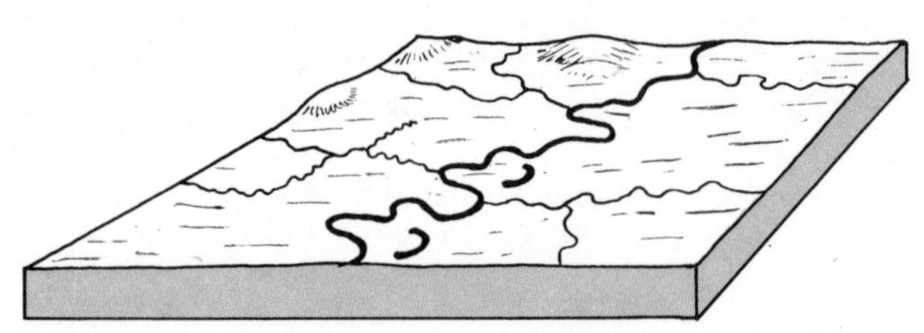

F. In old age, a peneplain with monadocks is formed

Fig 44 The cycle of erosion as proposed by W. M. Davis

which general statements should be based, supported by accurate measurements (quantification) at all stages.

Processes of landscape formation act upon the rocks of the earth's crust which vary both in their physical and chemical composition and, as a result of the upheavals of the geological past, in their arrangement. This is the **structure** of the land.

A third factor which must be considered is **time**, since any 'process' operating on a 'structure' will produce an ever changing landscape. It was this factor of a landscape constantly changing through time, which led W. M. Davis during the later years of the nineteenth century to propose an outline scheme (or 'model' in contemporary parlance) which saw the evolution of the landscape as an orderly sequence through stages which he called 'Youth', 'Maturity' and 'Old Age', roughly coinciding with the three sections of a river valley outlined at the beginning of this Study. *Fig 44* illustrates the stages of landscape development envisaged by Davis. This, the so-called **Cycle of Erosion,** was the first attempt to provide an overall pattern of thought and terminology for landscape studies and its influence upon Geomorphology is still strong. The details of the Cycle are explained fully in the references given at the end of this Study. Objections to the Davisian Cycle of Erosion are made in certain detailed aspects where Davis' assumptions can be shown to be incorrect and some writers today challenge the validity of his whole concept. There is also the danger that, in accepting uncritically the stages of Youth, Maturity, etc., students may limit their observation of the landscape to those features which fit the supposed pattern and ignore all other aspects.

Reading

*Chorley & Haggett: *Frontiers of Geographical Teaching*, Methuen 1967. Ch. 2, p. 21

Dury: *The Face of the Earth*, Ch. 6, pp. 61–74; Ch. 7, pp. 75–86

*Gresswell: *Physical Geography*, Ch. 17, pp. 174–198

*Holmes: *Principles of Physical Geology*, Ch. XVII, pp. 472–479; Ch. XIX, pp. 570–575

Horrocks: *Physical Geography and Climatology*, Ch. 5, pp. 74–75

Monkhouse: *Principles of Physical Geography*, Ch. 4, pp. 73–75

*Sparks: *Geomorphology*, Ch. 2, pp. 7–21; Ch. 4, pp. 55–74

Strahler: *Physical Geography*, Ch. 25, pp. 370–374

*Wooldridge & Morgan, *An Outline of Geomorphology* Ch. XII and Ch. XIII, pp. 140–172

Rakaia Gorge, South Island, New Zealand

RIVER TERRACES

Study A11

Fig 45 is an oblique aerial photograph of the Rakaia Gorge situated near the inland edge of the Canterbury Plains of South Island, New Zealand, near to the foot of the Southern Alps. Many interesting geomorphological features are shown (*see Further Work, Practical, 1*); the present Study concentrates on the features of the river valley. When studying an oblique aerial photograph the student should first attempt to establish the scale of the features shown. In the present instance a bridge crossing the river via the small island (centre foreground C5), the road from the bridge to the right foreground of the photograph (E7), and the dark lines of the trees planted as windbreaks between fields all help in determining scale. It may be noted in passing that in the clear air of New Zealand the mountains appear much closer than they really are.

The **braided** (or divided) course of the river in the foreground (A7) is indicative of the heavy load of silt which it carries; the general absence of vegetation colonizing the deposited material may well suggest that the river is seasonal in flow and that at times of spate, perhaps associated with spring melt water, the whole width of the bed may be full. From the foot of the mountains the land stretches towards the right of the photograph with a more or less horizontal surface, but the floor of the river valley is well below this level. Since the Canterbury Plains are formed of alluvial material of recent geological date, this fact in itself calls for explanation, for clearly the river at its present level could not have deposited material at the higher level. At some points the descent from the upper level of the Plains to the river level is by a single, steep step. In the left foreground (A6), for instance,

Fig 45 The Rakaia Gorge, South Island, New Zealand

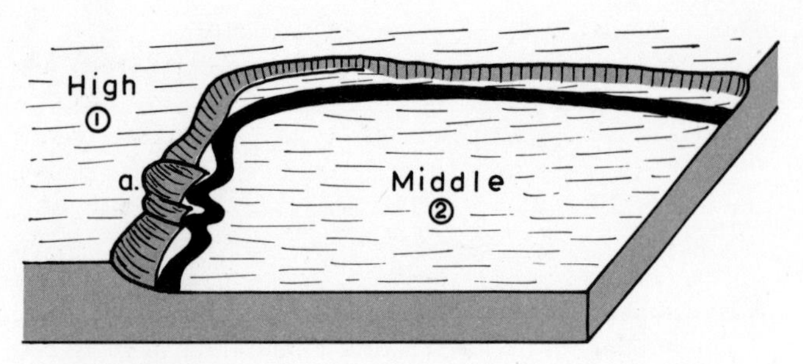

(a) First 'rejuvenation'. N.B. Formation of scalloped meander scar at 'a'

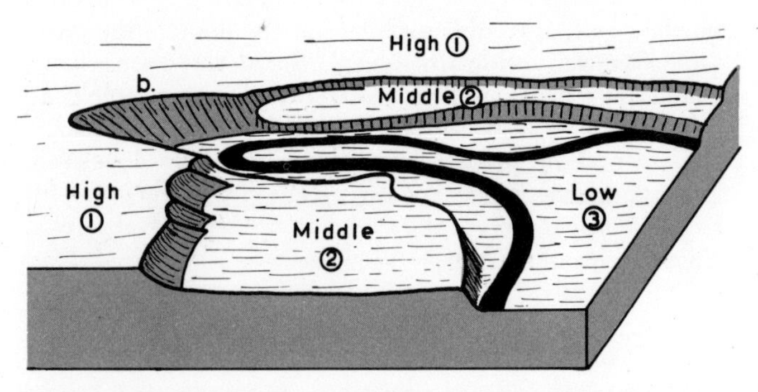

(b) Second 'rejuvenation'. Middle terrace has been dissected and at 'b' further erosion of High terrace has occurred

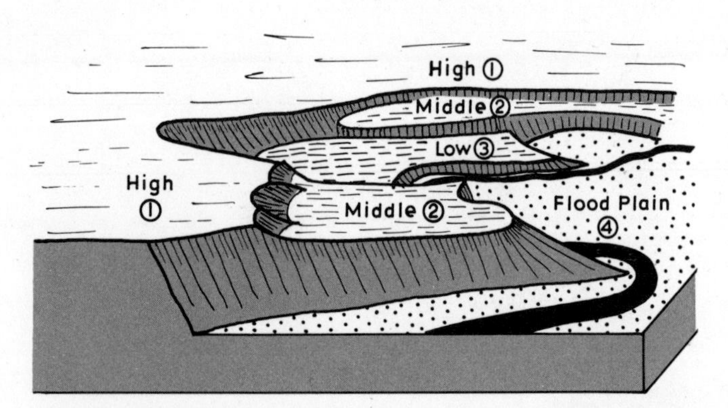

(c) Third 'rejuvenation'—portions of Low terrace removed and in foreground the river is eroding into both middle and high terraces

Fig 46

is a river cliff being actively eroded at its base, fretted by small **gullies** and rising to the level of the Plain. In many places there is evidence that the land is at present subject to active erosion, as for instance in steep sided, sharp edged minor valleys or gullies (e.g. D1) and the broken ground (F4) suggestive of slumping down towards the river.

Careful study of the photograph shows that in a number of places the drop from high Plain to river level is not in the form of a single step. The spur just right of centre (C1) shows clearly the existence of a number of smaller steps. Here, if one counts both the higher Plain surface and the present flood plain level there are four nearly horizontal 'treads' separated by three 'risers'. Since the upper surface is the level to which the original deposition of alluvium occurred and was therefore the river level of that period, and since the present river level is at the bottom of this flight of steps, it should be clear that the river has progressively cut into the original alluvial spread. It is not difficult to imagine each of the 'treads' as representing a level at which the river temporarily paused before renewing its downcutting. From this it will be clear that the more recently formed 'treads' or **terraces** are at a lower level than terraces of an earlier date. Such a conclusion can be substantiated by a detailed geological analysis of alluvium covering the terraces in this, as in many other examples.

It has been noted that in the left foreground (A6) there is a single step from the high Plain to the present river level. Here all the intervening terraces have been eroded away by the river at its present level. But on this same spur (at B3) a small step may be seen, not straight in plan but cuspate in shape. The level surface below this step must be an eroded remnant of an early terrace. Behind this spur (at B2) there is another curved cliff-like slope with a terrace at its base midway in elevation between the early terrace noted on the foreground spur and the present river level. Careful study of *Fig 46 a*, *b* and *c* together should make clear how these terrace remnants come into being. It may be useful to make tracings of these diagrams and superimpose them, or to redraw them from the photograph. The student will notice that while the downcutting or incision of the river is progressing there is ample time for the meander pattern of the river to change many times.

To sum up; we have in the photograph an area whose dominant land forms are of level plains separated by sharp **breaks of slope**. It is possible that such a pattern could be, at least in part, the result of erosion in an area of alternating resistant and less resistant horizontal strata. If such a landscape pattern is found in a region for which this explanation is not valid, it is likely that the explanation lies in the successive incision of the river, in stages, to lower and lower levels.

In the previous Study the Cycle of Erosion—the Davisian Cycle—was discussed and illustrated. Where breaks of slope and terraces are observed on the valley sides some modifications to this concept must be introduced to account for the irregularities. The significant point which emerges from the study of the Rakaia Valley is that a river can develop to the stage at which it deposits alluvium over a wide plain (the Old Age Stage of W. M. Davis) and at a later date acquire the increased energy which enables it to renew active erosion. Following the metaphor of Youth, Maturity and Old Age, this acquisition of renewed energy is termed **Rejuvenation.**

A number of environmental changes may produce the effect of rejuvenation. A change of climate, in particular increased precipitation in the river basin, will result in an increase in the volume of the river, and hence increase its energy both to erode vertically and to transport its load. The terraces of the Rakaia Valley can been shown to be the result of climatic changes: in this case each set of terraces converges at the level of terminal moraines, (*see Study C4*) and the rejuvenation can thus be associated positively with alternate advances and retreats of the glaciers from the Southern Alps, and with corresponding fluctuations in river volume. River Capture (*see Study A13*) will similarly increase the volume of the capturing river at the expense of its competitors. Again a stream may, in its head waters, start by eroding a vast quantity of unconsolidated glacial debris. As a result, such a stream will be heavily loaded so that broad spreads of alluvial material are deposited as soon as the gradient slackens. As time goes on the supply of ready fragmented material decreases; the load downstream is lessened and the river flowing across the alluvial plain uses less of its energy in transporting a load so that energy is available for vertical erosion.

All these changes in the load/discharge ratio of rivers may produce the effects of rejuvenation, but probably the most frequent cause of such features is a lowering of relative sea-level. There is ample evidence that such changes in relative sea-level have occurred: e.g. direct observation at the N end of the Gulf of Bothnia indicates that the land is at present rising from the sea at a rate of 30 cm every twenty-eight years. At points around the coast of the British Isles, and elsewhere, evidence can be found of beaches, sometimes backed by obvious marine cliffs, high above present-day sea-level, e.g. the 7.5 m **raised beach** of Western Scotland and that which lies 30–37.5 m up on the South Downs near Goodwood. Such changes can be **Eustatic**, that is, they are the result of a world wide lowering of sea-level consequent upon the imprisoning of vast quantities of water in ice-sheets during a glacial epoch (or are the result of pre-glacial eustatic changes); or they may be **isostatic**, resulting from local or regional uplift of the land. The latter is probably the most frequent cause of contemporary rejuvenation features since many areas of the world, including the British Isles, are still in process of adjusting their level having 'shaken free' of the enormous weight of the Pleistocene ice sheets. (*For amplification of these ideas see final section of Study C6, p. 92.*)

If one imagines the effect of a rapid fall in relative sea-level in a region where the land surface drops fairly steeply off shore, the sequence of events can be illustrated diagrammatically as follows:

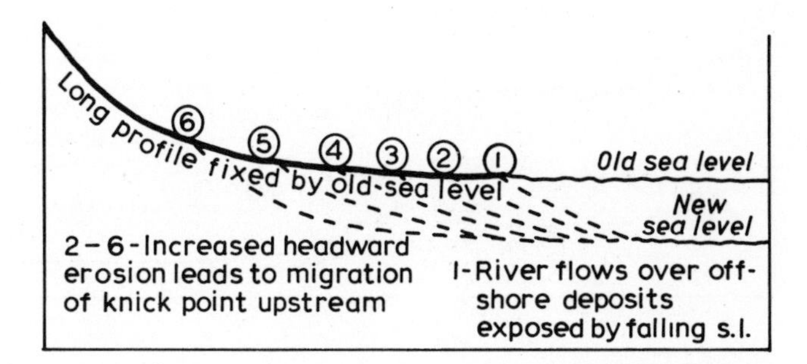

Fig 47 Upstream migration of knick points following rejuvenation

Initially the river falls from its old base-level to the new sea-level by a waterfall or rapids. The consequent increase in river energy leads to active headward erosion so that the lip of the falls retreats upstream and the change in gradient becomes less sharp. This change in gradient is the **knick point** or **head of rejuvenation.** Portions of the old valley floor remain uneroded to form terraces which converge at the knick point (*Fig 48*). A second lowering of sea-level would produce a second knick point and terraces at a lower level (*Fig 49*).

The number of cases in which relative sea-level falls on a coast with such a steep offshore slope as that envisaged above must be limited. More frequently a fall in sea-level will result in the exposure of a broad slope across which

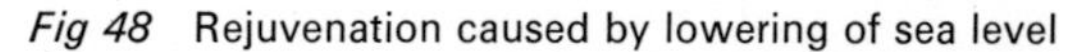

Fig 48 Rejuvenation caused by lowering of sea level

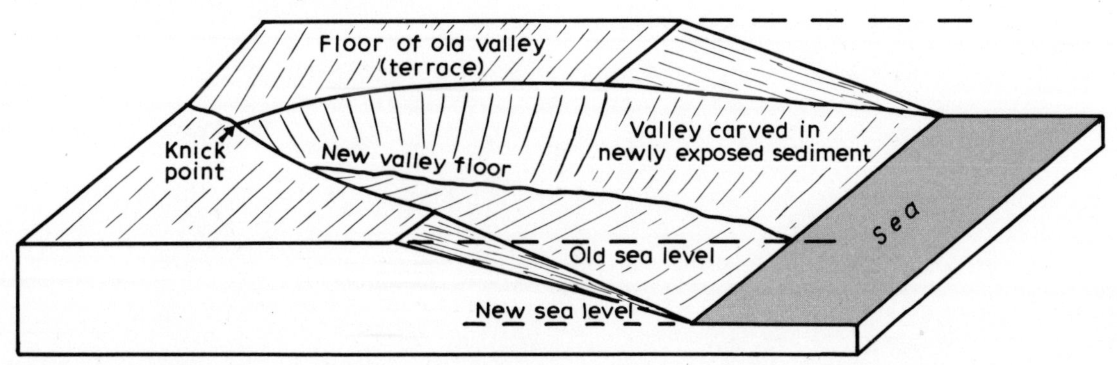

the river extends (*Fig 48*). In such cases the initial knick point will be less clearly defined, and becomes progressively less marked as it moves upstream, but the general principle and the mode of terrace formation remain unchanged. This is one reason why knick points are difficult to identify, even in the field, from the longitudinal profiles of rivers. A further difficulty is that a similar break in the slope of the long profile may be produced by the outcrop of a band of more resistant rock, and further it will be realized that the process of upstream migration of a knick point will be checked when a resistant outcrop is encountered and knick points of a later rejuvenation may 'catch up' until two or more such knick points coalesce.

River terraces, and other indications of rejuvenation which will be referred to in the next Study, are common features of the landscape of the British Isles. The fact that an area for this Study was selected from the other side of the world is an indication on the one hand of the universality of the rejuvenation process, and on the other of the difficulty of obtaining convincing photographs of these features which are often so small in scale or so disjointed in their occurrence that they cannot be clearly illustrated on one picture.

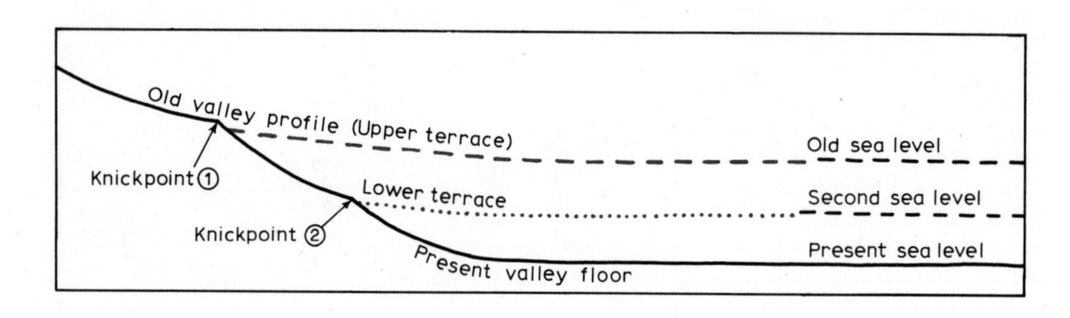

Fig 49 Long profile of river after two falls in sea level

FURTHER WORK

Practical

(1) Draw a sketch based on the photograph of the Rakaia Gorge (the student may find it better to make a tracing to provide the framework) and annotate it to show the physical features. Your annotation should include: braided stream; river cliff; low, middle and high terrace; meander scar; gulley; scree.

When later sections of this book have been studied you should be able to add: arête and corrie (*Study C2*); alluvial fan (*Study D2 and D3*).

(2) In the field: look for and accurately record any evidence of rejuvenation even if only of very local significance.

Reading

Anderson: *Splendour of Earth,* p. 291

Dury: *The Face of the Earth,* Ch. 7, pp. 75–87; Ch. 10, pp. 111–124

Gresswell: *Rivers and Valleys,* Ch. 9, pp. 93–97

——: *Beaches and Coastlines,* Ch. 7, pp. 105–111

——: *Physical Geography,* Ch. 18, pp. 205–216; Ch. 28, pp. 378–382

*Holmes: *Principles of Physical Geology,* Ch. XIX, pp. 576–587

Horrocks: *Physical Geography and Climatology,* Ch. 5, pp. 84–85, 88–89

Monkhouse: *Principles of Physical Geography,* Ch. 6, pp. 127–130

*Sparks: *Geomorphology,* Ch. 9, pp. 218–225

Strahler: *Physical Geography,* Ch. 24, pp. 357–358, Ex. 5, p. 368

Wooldridge & Morgan: *An Outline of Geomorphology,* Ch. XV, pp. 200–206

Study A12

The Lower Wye Valley

INCISED MEANDERS

O.S. 1: 63 360 (1.6 cm to 1 km *or* 1 in to 1 mile)
Sheet 155, Bristol and Newport

A superficial study of the meander forms of the R. Wye above Chepstow suggests a river in the 'Old Age' stage. Here are the apparently aimless curves and, at Tintern for example, the near formation of an ox-bow lake, or cut off—features which may be expected in association with a broad flood plain of greater width than the meander belt. But the contours tell a different story!

Even if one assumes that the flood plain extends as far as the lowest contour, the flat land bordering the river is nowhere more than 200 m wide, and this only at two or three points. In many places the river is immediately fringed by steep slopes, generally wooded, as at 541990 (Shorn Cliff); 530958 (Piercefield Cliffs); etc., etc. That these cliffs occur mainly against the outside of river curves is, of course, to be expected as was shown from the study of the Ribble Valley above Preston. But fairly steep slopes can also be found on the inside of river curves (537962), and land more than 60 m above river level (250 ft and 300 ft contours) occurs inside each bend, with steeper slopes on the N than the S side of each spur (*Fig 54*). A contour section drawn from spot height 771 (528976) to G.R. 550973 (*Fig 50*) shows that the right bank of the river (the W side of the valley) has a clearly marked break of slope at about 90 m (300 ft contour) with the steep slopes of Wynd Cliff rising above the more gently inclined spur within the meander curve.

Fig 50 Cross-section of Wye valley NE of St Arvans

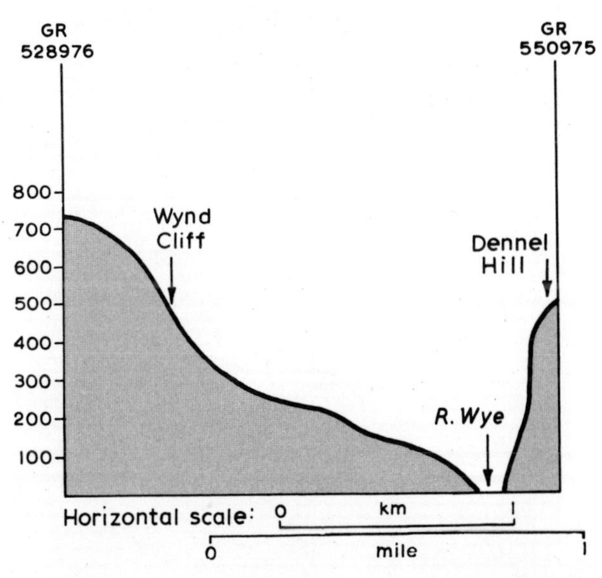

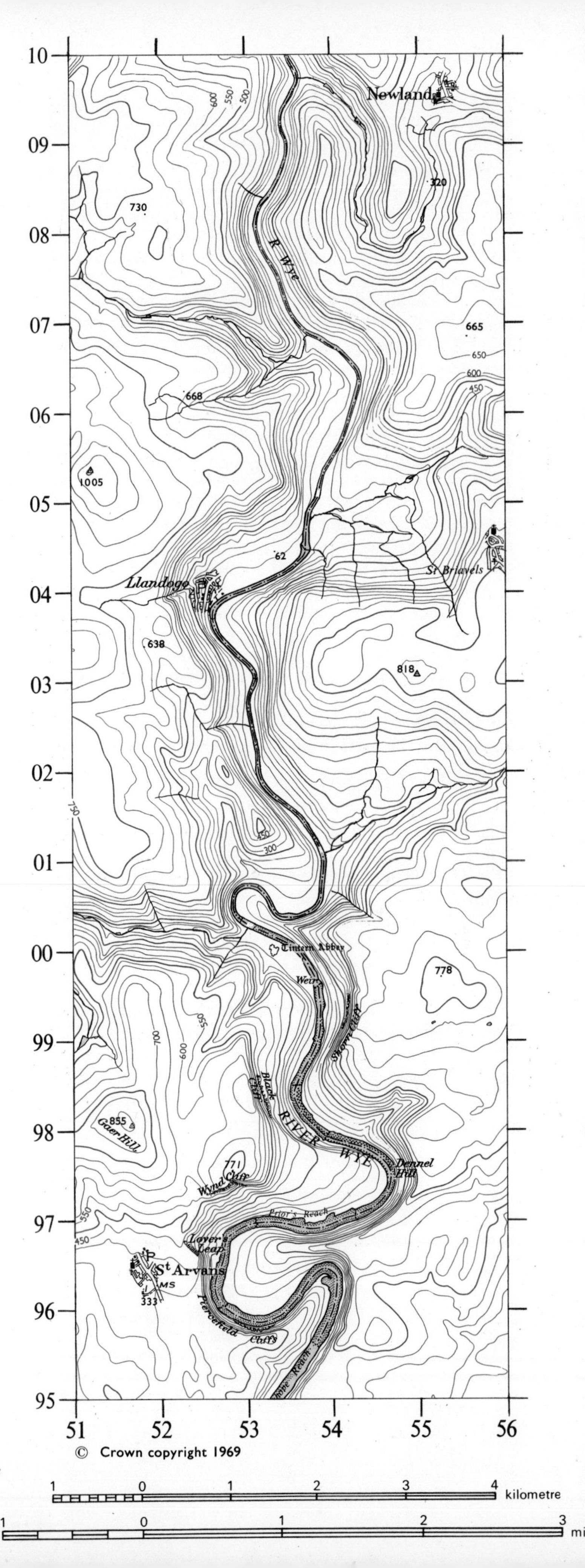

Fig 51 Wye Valley above Chepstow

Scale 1 : 63 360 (1·6 cm to 1 km *or* 1 in to 1 mile)

Contour interval 50 ft (15 m)

Exercise: Locate other points along the valley through which sections may be drawn to illustrate this dual nature of the slopes of the valley sides.

Well-marked breaks in slopes of this kind may reflect the outcrop of resistant rock strata on the valley sides; this can be checked against large-scale geological maps (*see Fig 52*). In the present instance, however, and where-ever geological differences do not occur, it is reasonable to suspect that the area has been moulded by two cycles of erosion. During the first of these cycles, sea-level was perhaps 175 m above its present level (i.e. at the level of the 250 ft contour), and at this time the R. Wye developed a flood plain up to a 1·6 km wide, across which it meandered much as the Ribble does today NE of Preston. After the Wye had attained its flood plain stage, the whole area was subjected to uplift relative to sea-level. The result of this elevation was to increase the river gradient and give it renewed erosional energy—a mature river was rejuvenated. While retaining its meandering form the R. Wye began to cut down into the older valley floor and produce the 'valley within a valley' form characteristic of rejuvenated landscapes (*Fig 53a–c*).

As the incision of the meanders into the earlier valley floor proceeded, the processes which bring about the downstream migration of river curves continued to operate. (*See Study A7.*) As a result one may see, on a greatly exaggerated scale, a slip off slope on the sides especially the S side of spurs, and steep river cliffs (undercut slopes) on the outside of the bends (*Fig 53c*). In cases where the uplift which led to rejuvenation has been very rapid, or where the rejuvenated streams flow through unresistant strata, the incision of meanders occurs so rapidly that meander migration does not have time to take effect. In such cases the river will flow in a sinuous trench, the sides of which are all of similar slope (*Fig 53d*). There are thus two kinds of **incised meanders** resulting from rejuvenation: (a) **ingrown**, in which stream migration has had time to develop slopes of varying steepness on the spur sides, and (b) **entrenched**, when rapid down-cutting produces a symmetrical trench.

Where the incision of meanders following rejuvenation is accompanied by migration of meanders, it is likely that at some stage a meander loop will become cut off by the same processes which produce ox-bow lakes on a flood plain. After such a loop has been abandoned the main stream continues to cut down so that the abandoned section of the valley is left at the level it occupied when the 'cut-off' occurred, possibly modified by erosion by a

Fig 52 Wye Valley—geology

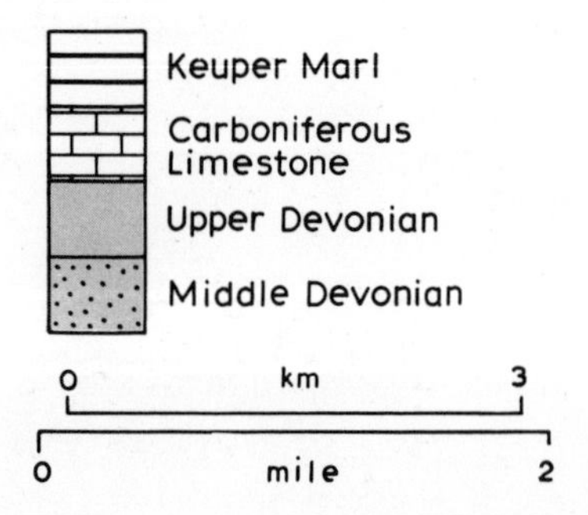

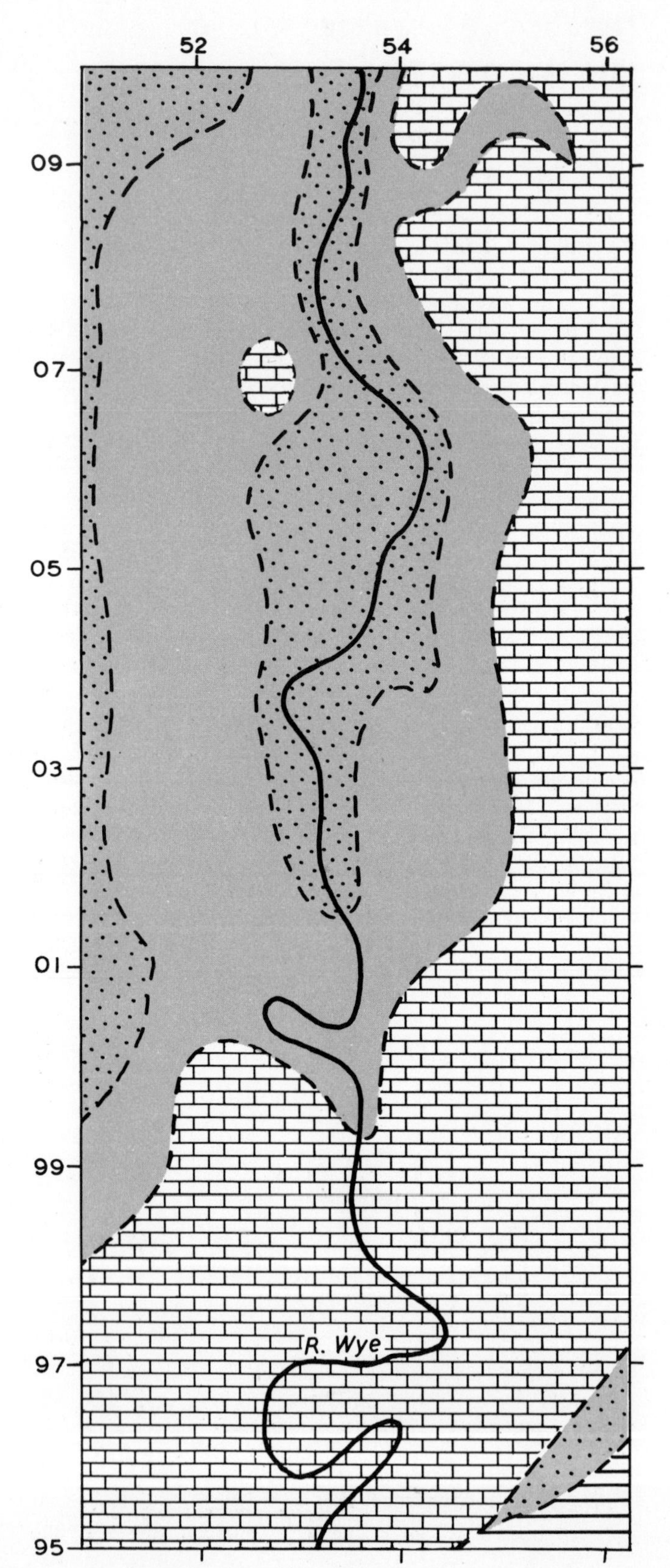

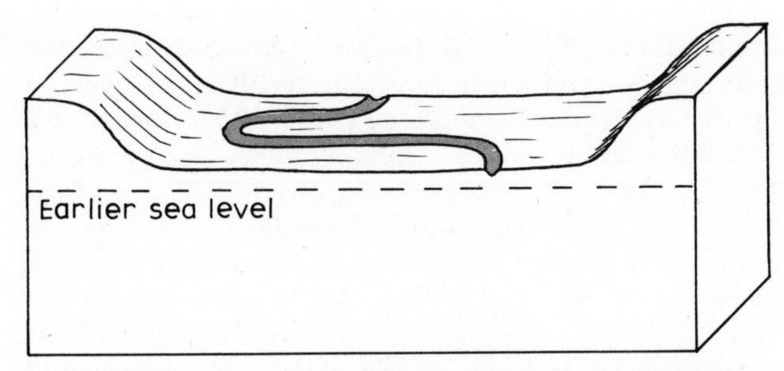

(a) *Stage I*. River meandering across flood plain only a little above sea level

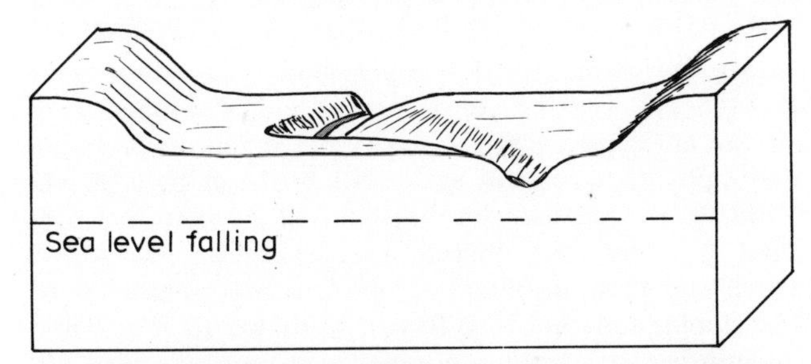

(b) *Stage II*. Sea level falling (or land rising). River is 'rejuvenated' and begins to deepen its valley within earlier flood plain

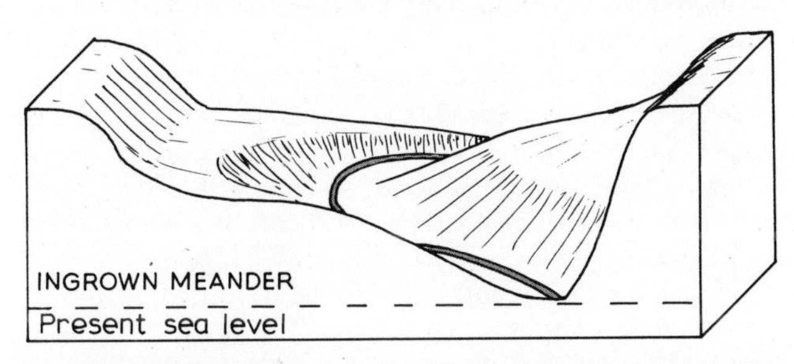

(c) *Stage III*. River erosion continues as sea level falls. Meanders now deeply incised—most remnants of flood plain removed. Downstream migration of meanders during incision produces asymmetrical cross-section

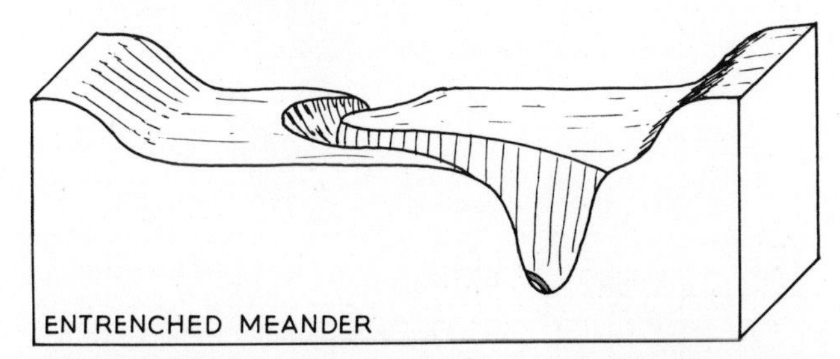

(d) Rapid downcutting by river with no time for meander migration produces symmetrical cross-section—more of original flood plain remains uneroded

Fig 53

tributary stream for which it provided an easy route. In the case of the Wye Valley area shown, two such abandoned loops may be identified—one at Newland in the far N of the extract (around 5408) and one W of St. Briavel (around 5404). The latter is only a little above present river level and must therefore have been formed much more recently than the N example.

Besides being rejuvenated, the R. Wye also provides an example of **superimposed drainage**. The courses of the river and its tributaries show virtually no adjustment to the geological outcrops shown on the present-day map (*Fig 52*). Where such a situation exists it is reasonable to suspect that the river system evolved on a cover of more recent rocks which have subsequently been removed by erosion exposing older rocks with a quite different structural pattern. The surface separating older from newer rocks in such cases is an **unconformity** (*Fig 55*).

This short study of a section of the Wye Valley has thus introduced a number of factors which may modify the land forms associated with river valleys and has shown that an understanding of the present-day landscape may have to be sought far back in the geological history of the area. Indeed in the British Isles this kind of complexity is more frequently encountered than the simple pattern presented in the preceding Studies. It is frequently the case that the main evidence of past changes in sea-level is now to be found only in the remnants of much earlier plains forming the summits of hills, or in unexpected

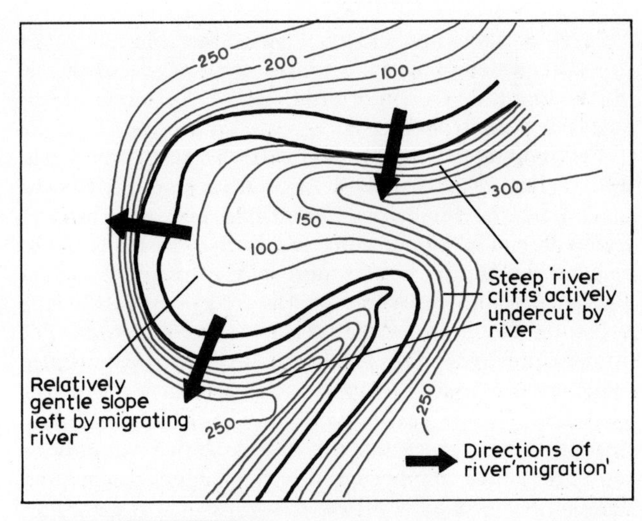

Fig 54 Sketch map of spur east of St Arvans (GR 5396) showing features of 'ingrown' meander

Fig 55 One type of superimposed drainage

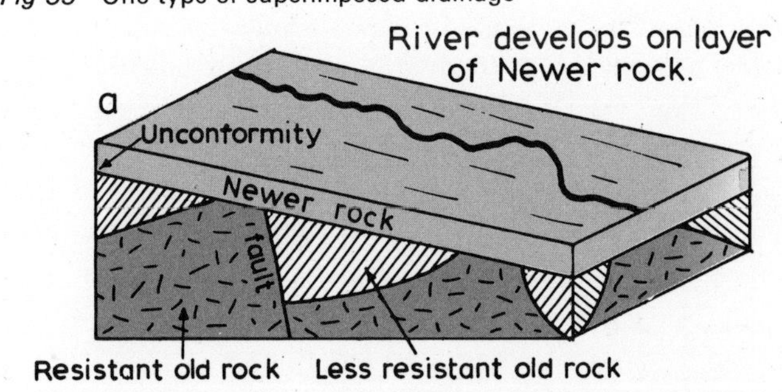

breaks of slope on valley sides. Such remnants are broadly termed **planation surfaces** and their analysis and elucidation are an important part of the work of the research geomorphologists.

Frequent uplift of the land since Tertiary times, the many changes in sea-level associated with Pleistocene glaciation, the climatic modifications, and interruptions to the fluvial erosional processes during the Ice Ages, as well as the complex structural history of much of the British Isles, all contribute to the frequency with which evidence of long-past events occurs on our landscape. The student must therefore beware of stating over-simplified explanations based on only a small fraction of the evidence. This is particularly true in the case of map interpretation, where any conclusions drawn can only be tentative until supported by field observation and detailed study of all available data.

FURTHER WORK

Practical

(1) Reference has already been made (*Study A2*) to the significance of the angle of slope in geomorphological studies. The mapping of slope angles is a useful exercise which can be applied to any area under study.

First the relationship between the spacing of contours on the map and the angle of slope must be established. This can be calculated mathematically, or a very close approximation may be reached by constructing a diagram on graph paper with $\frac{1}{10}$ in squares. To obtain a reasonable degree of accuracy it is necessary to enlarge both horizontal and vertical scales by ten. A horizontal base line of 5 in will thus represent $\frac{1}{2}$ mile (1 in to 528 ft) and $\frac{1}{10}$ in on the vertical scale represents 52.8 ft. Bearing in mind the limits of accuracy in drawing, this may be considered as 50 ft. Lines are next constructed radiating from one end of the base line at 5° intervals up to 30° (*Fig 56*). It will then be seen that the 5° line, for example, reaches the line representing 50 ft in height in a horizontal distance of 1.1 in; the 100 ft line in 2.2 in; 150 ft in 3.3 in etc., i.e. a spacing of 1.1 in between 50 ft levels. Since the scale of our diagram was exaggerated ten times, it is clear that a slope of 5° on the ground will be shown on a 1 in to 1 mile map by contours approximately 0.1 in apart. In the same way, spacing of contours can be worked out for slopes of 10°, 15°, etc. and it will be seen, for example, that a slope of 25° will be indicated by six contours to $\frac{1}{10}$ in.

If metric graph paper is used, the same method can be applied using the following data: The map scale of 1:63 360 (1.6 cm to 1 km) is enlarged ten times for the horizontal scale, so that 16 cm represent 1 km. The equivalent vertical scale will be 1 cm to 6336 cm or 63.36 m. Because the map contours are in feet this must be converted, therefore 1 cm represents 193 ft which can be approximated to 1 cm represents 200 ft. The sloping lines are then drawn as shown on *Fig 56* and the numbers of contours per centimetre may be read off.

The graph which has been drawn should be kept for further reference, and a similar diagram prepared for use with 1:25 000 maps.

In using slope angle as a means of describing relief it must always be realized that hillsides appear much steeper to the observer at ground level than an angle drawn on paper would suggest, so that it is very important to link theory with experience by measuring slope angles in the field as suggested in Study A2.

Using the data on contour spacing obtained as above, a slope angle map of a part of the Wye Valley has been drawn (*Fig 57*) and, for comparison, a similar map of part of the Ribble Valley (*Study A7*). To make the task of drawing such maps as simple as possible, the examples use only three slope categories:

(a) steeper than 25°—i.e. more than five contours to $\frac{1}{10}$ in (or 25 contours per centimetre)

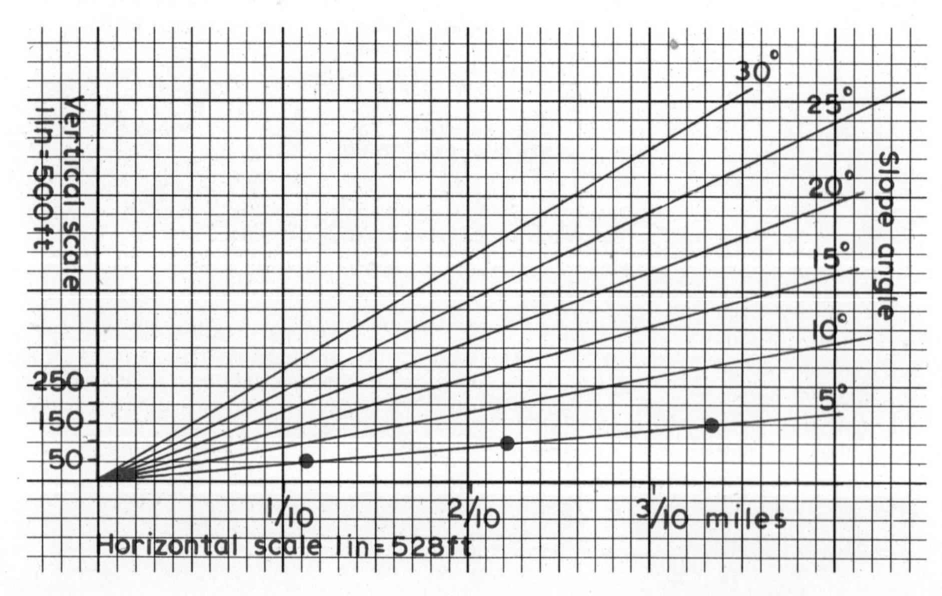

Fig 56 Diagram to calculate contour spacing in relation to slope angle on a 1 in to 1 mile O.S. map. (N.B. Vertical exaggeration is negligible)

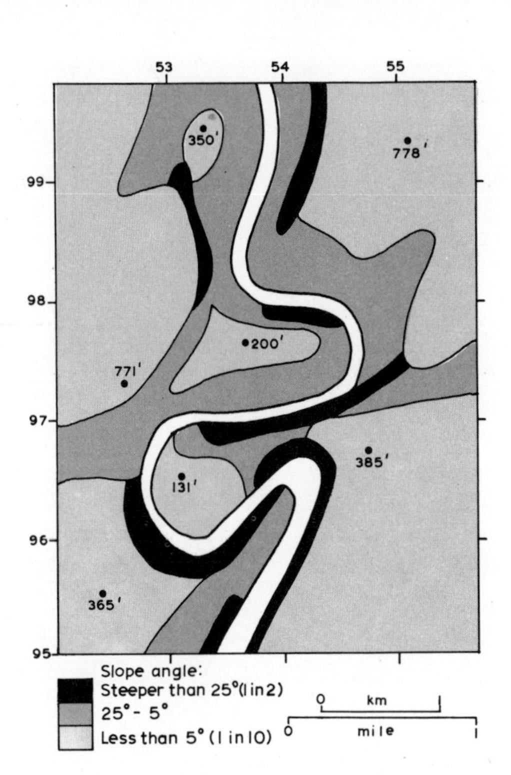

Fig 57 Slope angle map of part of Wye Valley

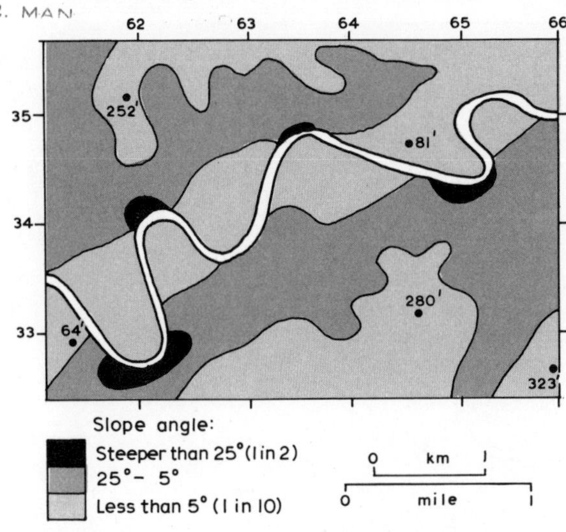

Fig 58 Slope angle map of part of Ribble Valley

(b) less than 5°—i.e. contours more than $\frac{1}{10}$ in apart (less than 4 contours per centimetre)
(c) all intermediate slopes between (a) and (b).

Such maps are, of course, full of inaccuracies, and no precise conclusions can be reached from them. Nevertheless they are an interesting method of presenting morphological data which students may find helpful and the method has the great merit that the drawing of such maps calls for very close study of the contours of an area whereas the construction of contour sections concentrates attention along a narrow, and possibly untypical, strip. In the present examples there is a clear difference in the slope patterns of the Wye and the Ribble which calls for explanation and any process of map analysis which raises questions in the student's mind is of the greatest value.

(2) With reference to a suitable geological map (the 10 mile to 1 in map will suffice) and the 1:63 360 Sheet 155 draw a simplified sketch map to show the relationship of the Bristol Avon, the R. Frome and Colliter's Brook (560699) to the outcrop of Keuper Marl and Carboniferous Limestone. Use your sketch map in conjunction with the paragraph on superimposed drainage (*above p. 35*) to suggest a possible explanation of the formation of Clifton Gorge (G.R. 5673).

(3) If the *Geological Journal* of September 1955 (Vol. 121(3)) is available, study the paper on Terraces of the Salisbury Avon in conjunction with the O.S. maps of the area—1:63 360, Sheet 179 (Bournemouth) or 1:25 000 SZ/19 and SU/10. Try to find out examples of any similar studies and especially whether any have been made near your home area.

Reading

Dury: *The Face of the Earth*, Ch. 7, pp. 75–87; Superimposed drainage—Ch. 3, p. 25
Gresswell: *Physical Geography*, Ch. 18, pp. 205–216.
Hardy & Monkhouse: *Physical Landscape in Pictures*, p. 50
Holmes: *Principles of Physical Geology*, Ch. XIX, pp. 588–591; Superimposed drainage—pp. 563–566
Horrocks: *Physical Geography and Climatology*, Ch. 5, pp. 79–80; Superimposed drainage—pp. 85–87
Monkhouse: *Principles of Physical Geography*, Ch. 6, pp. 130–131; Superimposed drainage—Ch. 6, pp. 151–153
——: *Landscape from the Air*, p. 23
Scovel: *Atlas of Landforms*, pp. 84–5, 90–1, 94–5
*Sparks: *Geomorphology*, Ch. 8, pp. 225–236; Superimposed drainage—Ch. 6, pp. 118–133
Strahler: *Physical Geography*, Ch. 24, pp. 359–363; Ex. 4, p. 367
Trueman: *Geology and Scenery*, Ch. 18, pp. 256–258; Ch. 19, pp. 259–267
Wooldridge & Morgan: *An Outline of Geomorphology*, Ch. XV, pp. 206–210

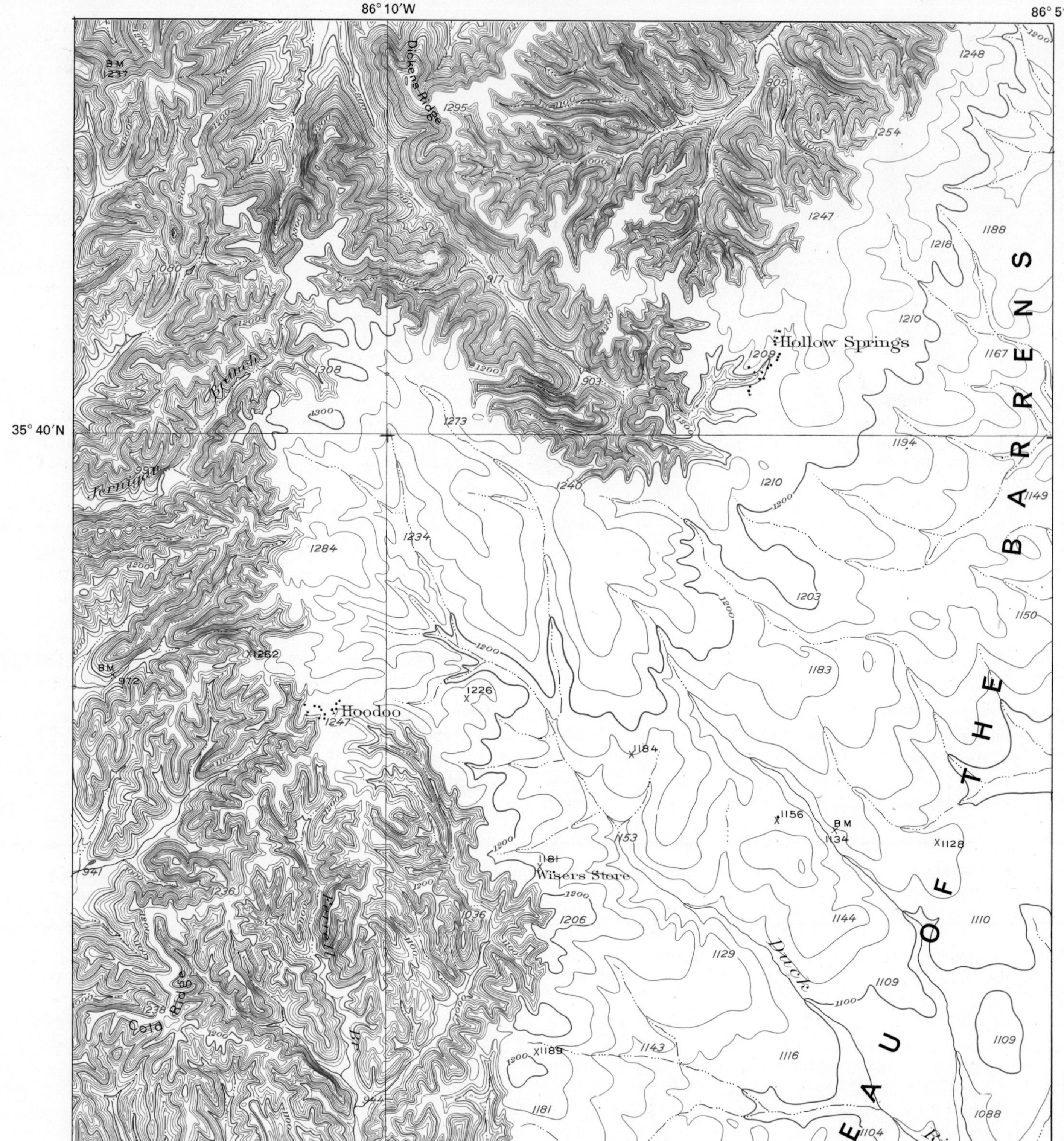

86° 10′W

86° 5′W

35° 40′N

35° 35′N

Dickens Ridge

Hollow Springs

Hoodoo

Wisers Store

Cold Ridge

Duck River

PLATEAU OF THE BARRENS

Fig 59 Hollow Springs, Tennessee, U.S.A.

Scale 1 : 62 500 (1.7 cm to 1 km *or* 1 inch to 1 mile)

Contour interval 20 feet (6 m)

1 0 1 kilometre

1 ½ 0 1 mile

Hollow Springs, Tennessee, U.S.A.

RIVER CAPTURE

U.S. Geological Survey 1: 62 500 (about 1.7 cm to 1 km *or* 1 in to 1 mile)
Hollow Springs Quadrilateral

The original 1:62 500 U.S. Geological Survey map from which *Fig 59* was prepared provides an excellent lesson of the dangers of jumping to conclusions about relief features shown on a map. Almost every student making a first appraisal of the map (which has a 20 ft contour interval) assumes high land on the west falling steeply to the plain of Duck River. Study of contour values and spot heights reveals that the reverse is true. The map shows a plateau surface in the east with only very gradual slopes, varying in altitude from about 320 m (spot height 1069 ft) at the SE corner of the map to 400 m (1300 ft contour) on its W side, whence it descends sharply in a mass of broken country to about 280 m (spot height 941 ft) at the W edge of the map.

The map shows therefore a plateau being actively dissected on its W edge, and moreover there is clear indication that the area contains landscapes in two contrasting stages of evolution. On the plateau the drainage is to the S and E in broad, shallow valleys only some 25 m lower than the interfluves, and the Duck River in the S of the map flows in a valley with very gently sloping sides. By contrast the streams dissecting the edge of the plateau exhibit the steep sided, narrow valleys and the steep gradients of a youthful landscape.

The Hollow Springs area thus exhibits one set of circumstances in which the process known as **river capture** is likely to occur: namely the impingement of a stream with considerable erosive power on to the catchment area of a less energetic river system. The cause of such a situation is likely to be that rejuvenation knick points have moved more rapidly along one of the river systems than along the other. A generalized cross-section E–W across the map extract (*Fig 60*) shows this situation. The steep W gradients will give the west flowing streams great erosive power, some of which will be utilized in rapid headward erosion thus pushing the divide further E diverting or 'capturing' bit by bit the surface water which at present flows across the plateau (*Fig 61*).

As this happens gradually it is common to find that the course of a successful 'attacking' river perpetuates some of the pattern of its 'victim'. *Fig 62* shows the rivers in the area of the settlement of Hollow Springs, and it can be seen that some of the tributaries of the 'attacking' river appear to flow from the reverse direction from that which might be expected. These tributaries (A–B and C–D) perpetuate the courses of streams which once flowed E across the plateau at this point. A sharp river bend formed in this way is known as an **elbow of capture**.

The essential condition for river capture is that one river, or river system, shall have greater energy for downcutting and headward erosion than a neighbouring stream. This condition may be the result of rejuvenation affecting one river before its neighbour due to relative distance from the sea; it may be due to climatic factors leading to heavier rainfall in one catchment area; or it may result from the fact that the capturing stream is cutting back rapidly into less resistant rocks than its rival. These factors may occur in combination.

The elucidation of the development of river capture demands painstaking research on the ground and cannot

Fig 61 One case in which river capture occurs

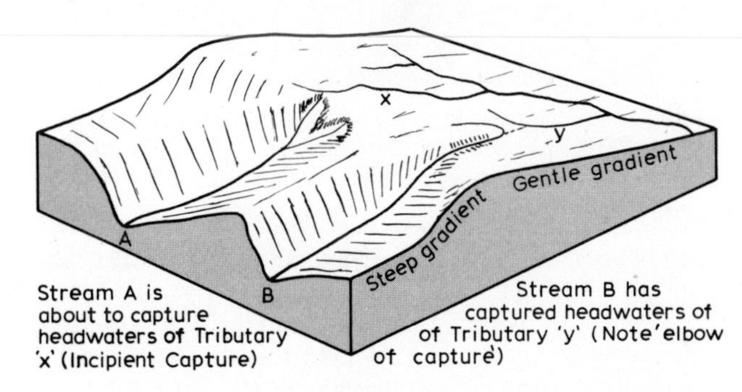

Fig 60 Generalized section E–W across Hollow Springs map extract

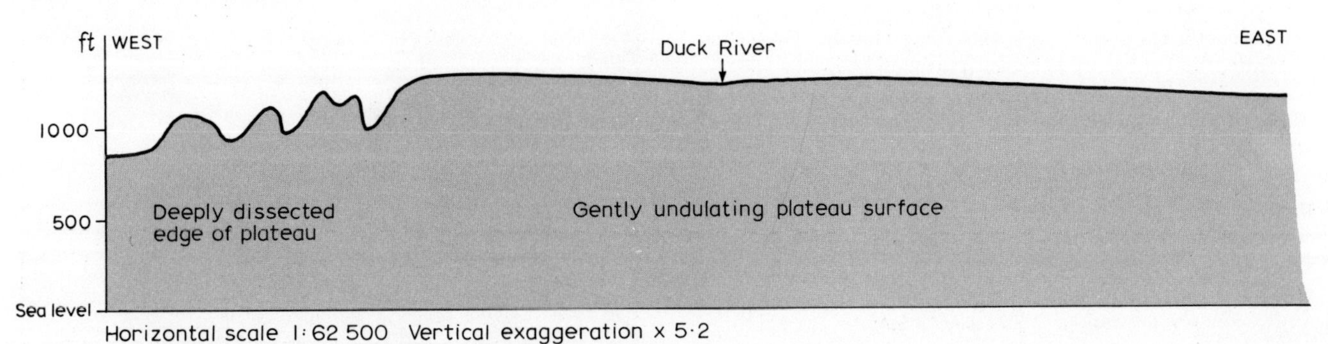

Fig 62 'Elbows of Capture' near Hollow Springs, Tennessee. Note that valleys A–B and C–D perpetuate pre-capture orientation of streams and are in reverse direction to capturing river

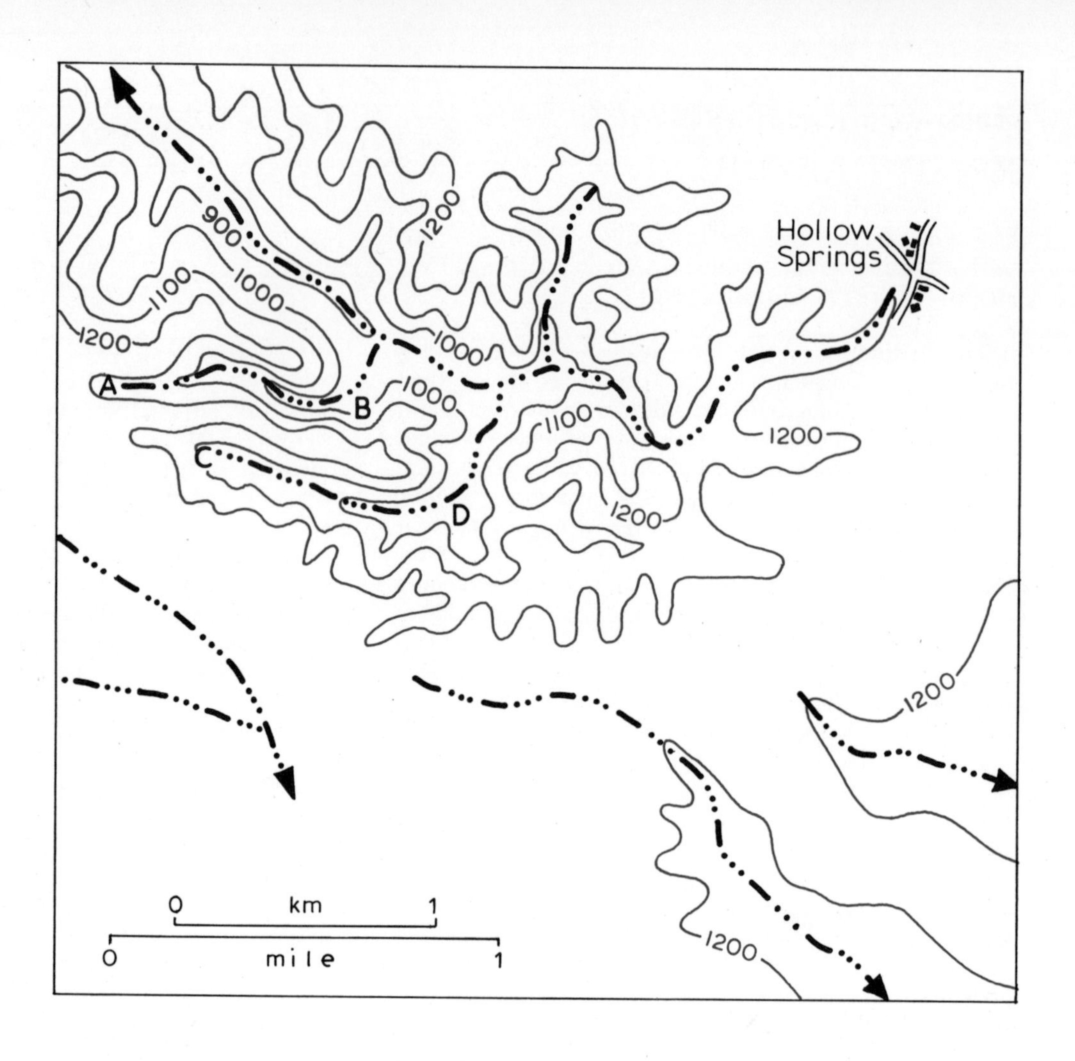

usually be recognized with any certainty from topographical maps. The following example should be studied in conjunction with the 1:63 360 or 1:250 000 ("Quarter Inch") map:

Yorkshire Ouse (1:63 360, Sheets 91 and 97; 1:250 000, Sheets 9 and 11)

The Swale, Ure, Nidd, Wharfe, Aire and Calder (*Fig 63*) in their upper courses are all **consequent** streams—that is to say their E direction of flow is a direct consequence of the E dip of the rock strata on this side of the Pennine anticline. The original courses of these rivers would almost certainly have carried their waters directly and partially independently to the sea. The R. Ouse has extended its course by headward erosion in the relatively less resistant Triassic deposits, capturing in turn the Nidd, the Ure and the Swale. A tributary stream such as the Ouse which cuts its valley along a less resistant outcrop and is thus flowing parallel to the **strike** of the rocks is known as a **subsequent** stream. The nomenclature 'consequent' and 'subsequent,' although explained in this section for convenience, is used whether or not river capture has occurred. (*See also Study E6, especially Fig 183.*)

Fig 63 Rivers of Yorkshire, showing capture by Ouse along outcrop of 'soft' Triassic beds (dotted)

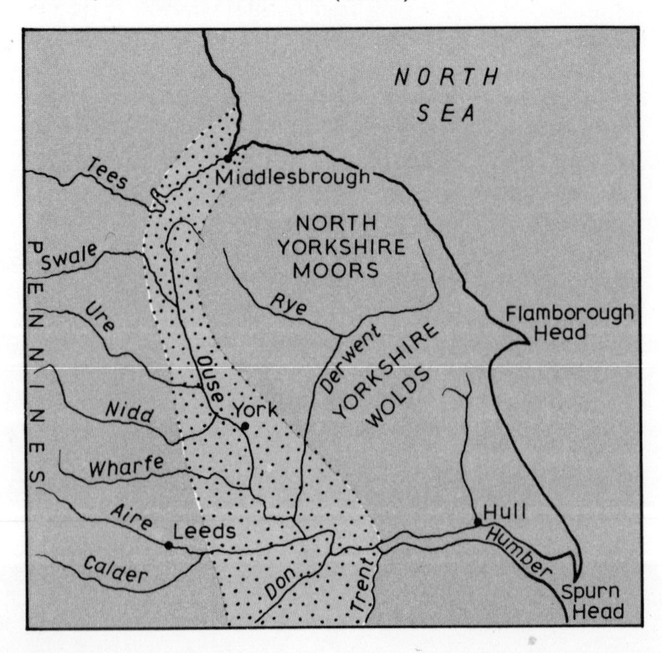

FURTHER WORK

Practical

(1) Textbooks in Geomorphology should be treated as a guide to, and a poor substitute for, field work. Such field study is not limited to a few 'fortunate' locations in which features described in books are to be found and identified. In any part of the British Isles, even in cities where minor slopes can be observed, it is possible to make valid studies. Neither does valid field work consist of the student being taken to a site, told what to look at and faithfully noting and remembering what has been pointed out. A student working on his own can achieve a greater understanding of the subject by applying the following programme in *any* area in which he finds himself.

First **observe** in the field:
- where are the slopes located?
- how steep are they?
- how are the slopes related to each other?
- where are there sizeable level surfaces?
- are there any marked 'breaks of slope' (i.e. sharp changes in the angle of slope)?
- are there few or many rivers?
- what is the size, gradient, velocity, etc. of the rivers?

and so on. Supplement findings by the study of maps and/or photographs until a more or less complete **description** can be prepared in the form of sketch maps, field sketches, diagrams and notes.

Then attempt to **analyse** what has been described:
- are the features being formed by processes acting at the present time, e.g. river cliff, flood plain?
- if so, how are the processes operating to produce the landform?
- is the landscape changing quickly or slowly?
- can the observed features be related to the geological structure? (This theme is developed further in *Section E*.)
- if the feature cannot be explained in terms of present-day processes or geological structure, can it be due to some condition which has occurred in the area in the past, e.g. a different climate, such as an Ice Age (*see Section C*), or a different sea-level?

Remember that it is at least as important to ask the right questions and to think out ideas for yourself as it is to get the right answers.

(2) Use *Fig 64*, if possible in conjunction with O.S. Sheet 112, 1:63 360 (Nottingham) or Sheet 11, 1:250 000, to explain how a process of river capture may account for the present drainage pattern of this part of the Trent Basin.

(3) With reference to the Hollow Springs map extract:

(a) Explain, with a sketch map, what will happen as the S flowing stream W of Wisers Store extends its course by headward erosion.

(b) Draw cross-sections to illustrate the characteristic features of the Duck River and of one river on the W of the map extract.

(c) Use the indications of latitude and longitude to locate this area on a good atlas map of the U.S.A. Can you see how the rivers shown eventually find their way to the sea?

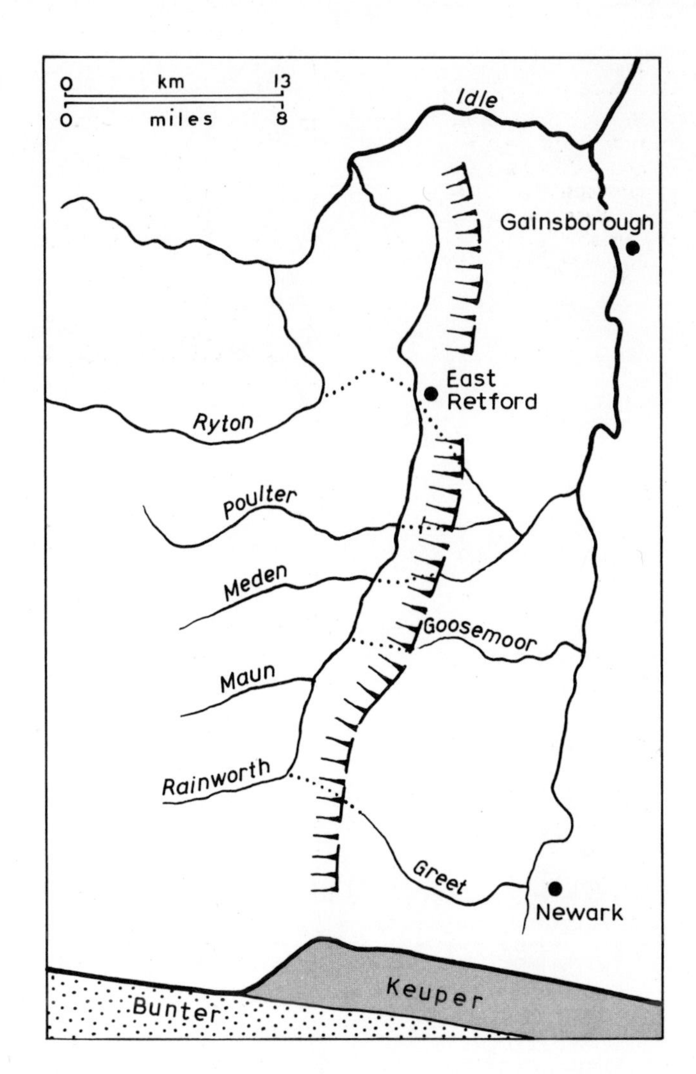

Fig 64 River capture in Nottinghamshire (after Gresswell)

Reading

Carter: *Land Forms and Life*, Section 9, pp. 69–75
Dury: *The Face of the Earth*, Ch. 3 pp. 21–23; Ch. 17, pp. 207–209, 212–215
Gresswell: *Rivers and Valleys*, Ch. 10, pp. 115–122
——: *Physical Geography*, Ch. 17, pp. 176–182; Ch. 18, pp. 211–212
Holmes: *Principles of Physical Geology*, Ch. XIX, pp. 558–563
Horrocks: *Physical Geography and Climatology*, Ch. 5, pp. 80–84.
Monkhouse: *Principles of Physical Geography*, Ch. 6, pp. 142–144
Scovel: *Atlas of Landforms*, p. 96
*Sparks: *Geomorphology*, Ch. 6, pp. 101–118
Wooldridge & Morgan: *An Outline of Geomorphology*, Ch. XIV, pp. 175–179

Study B1

WAVES—AN INTRODUCTION TO COASTAL STUDIES

Mention has been made already (*Study A7, p. 20*) of the fact that water moving over a solid surface begins to oscillate in such a way that meanders develop, and it was pointed out that this process is akin to the movement of air over water producing waves. The physical processes involved in these movements are complex. The geomorphologist's interest is with the resultant meanders and waves, rather than the physics of their formation. But if one is to understand how coastal land forms arise, it is necessary to have a clear picture of some aspects of wave motion.

The familiar breaking wave of our beaches is, in fact, a special type of wave which will be considered below. The **wave of oscillation** which occurs in the open sea is more widespread and is the type of wave initially produced by the action of wind over water.

This wave of the open sea is *not* a moving body of water; rather it is a forward surge of energy transferred from the atmosphere to the sea by friction with the wind.

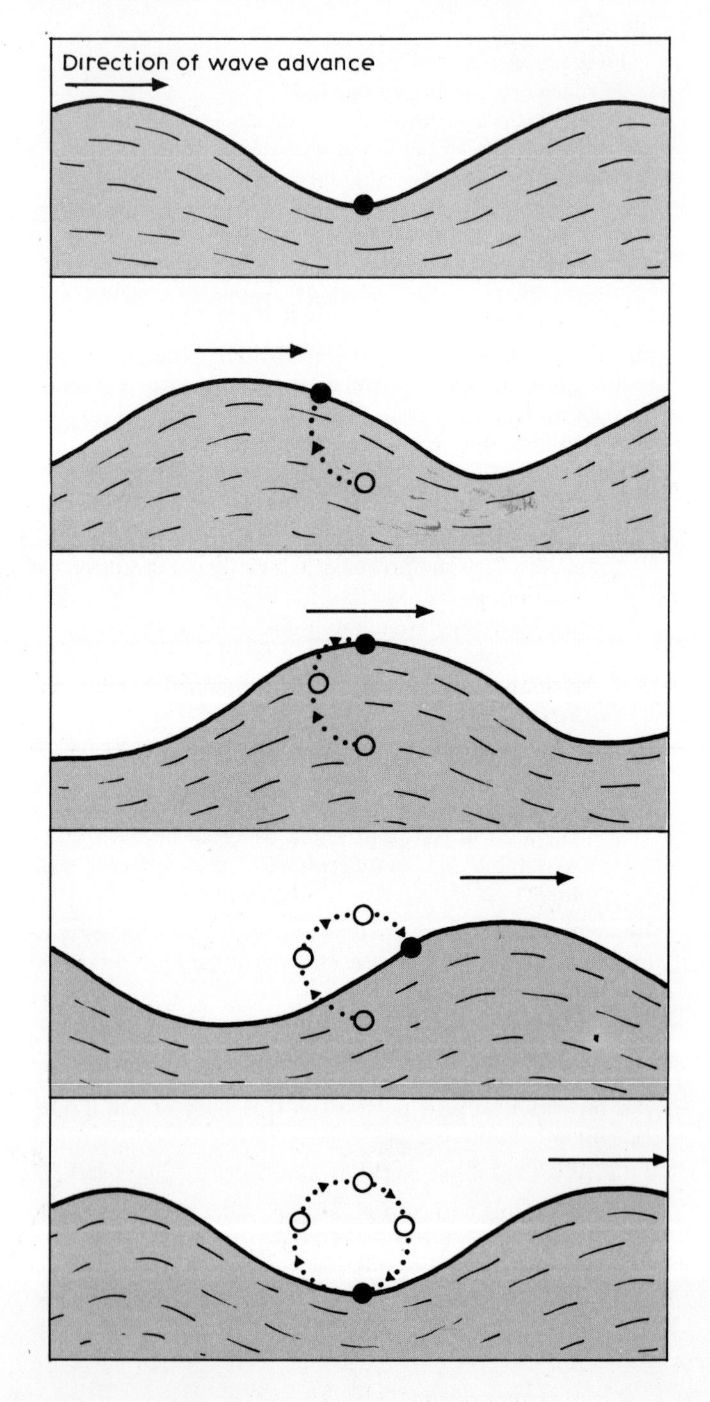

Fig 65 Motion of floating ball as a wave passes
Previous positions of ball shown by open circles

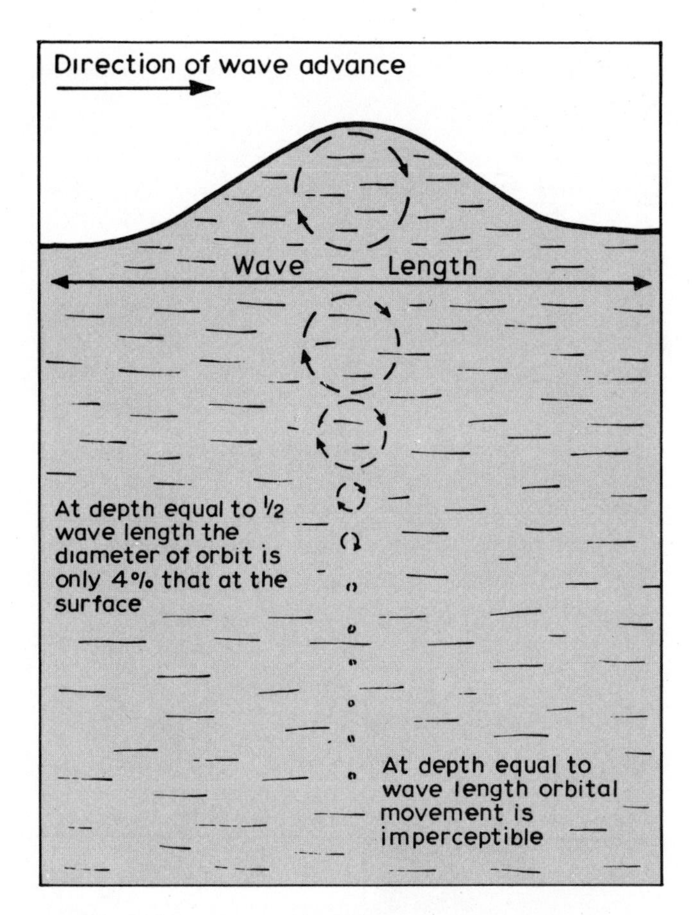

Fig 66 Orbital movement in waves decreases with depth

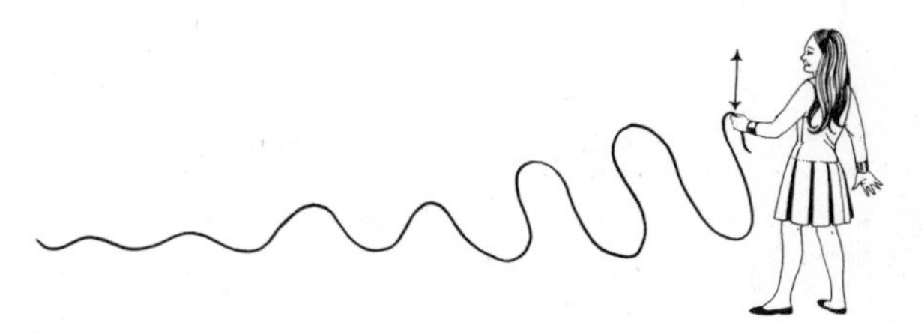

Fig 67 Waves formed by snaking a rope

Useful analogies may be found, firstly when a rope is fixed at one end and 'snaked' (*Fig 67*). Ripples move along the rope, but the rope itself does not move forward. Similarly a cornfield on a windy day appears to have waves moving across it, though obviously each plant remains rooted in the same place. If the movement of a ball floating on waves can be observed in deep water—e.g. at the end of a pier—it will be seen that as each wave passes, the ball moves backwards and then upwards as the wave trough passes and then forwards and downwards on the succeeding crest (*Fig 65*). In other words the ball describes a circular orbit in a vertical plane, returning to the same spot after each wave. This movement reflects the movements of water particles in a wave—each particle moves in a roughly circular orbit, the diameter of which decreases with depth until at a depth approximately equal to the wave length there is no movement which can be attributed to the passage of waves (*Fig 66*).

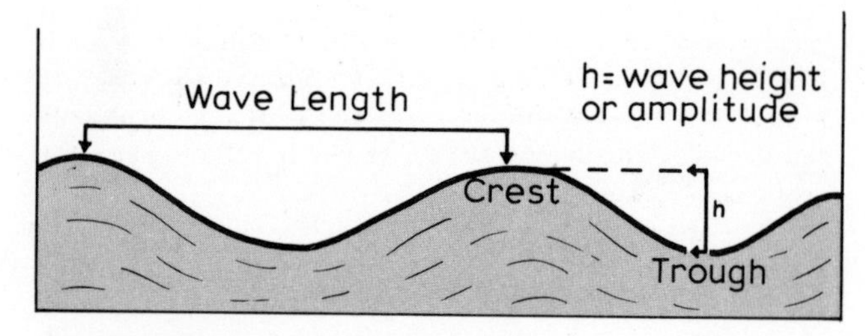

Fig 68 Terms used to describe waves

The **height**, or **amplitude** of waves, measured from trough to crest (*Fig 68*) is dependent firstly on the strength of the wind and secondly on the distance over which the wind has been forming the wave. On even a small pond it can be observed that the wavelets which are very small on the windward side become larger as they cross the pond, reaching a maximum height on the side furthest from the direction of wind approach (*Fig 69*).

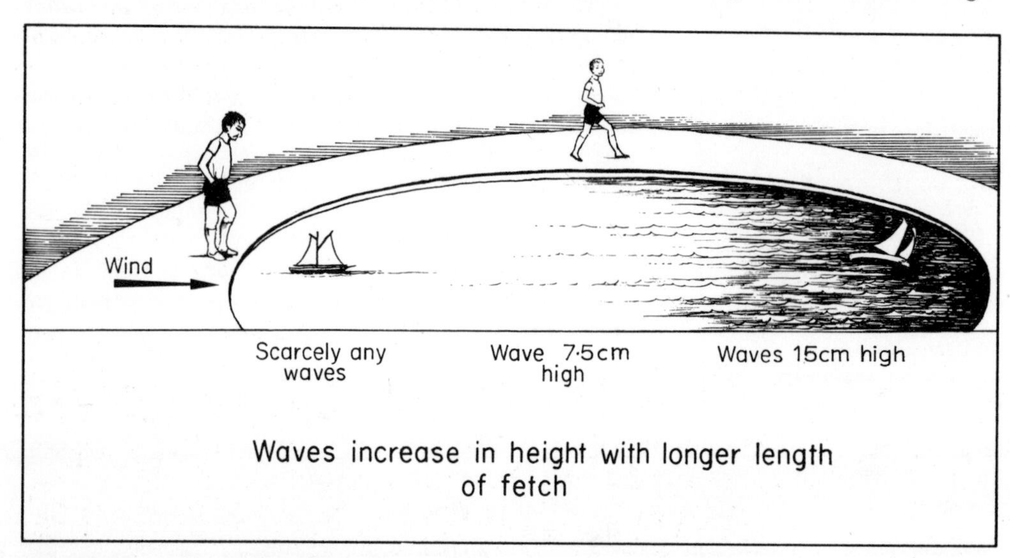

Fig 69 Waves on a pond

This distance over which waves are driven by the wind is the **fetch**—generally speaking the longer the fetch, the greater will be the strength of the waves. Applied to the coastline this means that the most powerful waves will reach places where the uninterrupted fetch is greatest; in the case of the British Isles those W facing coasts unprotected from the Atlantic by any land will receive the strongest waves. Conversely in the case of coasts facing across relatively narrow seas such as the English Channel or the Irish Sea the fetch is insufficient for the generation of really large waves.

Once formed, large waves will continue to move across the ocean for considerable distances beyond the area in which they were formed. **Free waves** of this type reduce in amplitude and increase in wave length, reaching a distant shore line as a **swell**. An observer on a coast may thus observe a complex of varying waves at any one time. Some breakers may represent free waves, generated far away and with a long wave length (say 300 m), arriving every 15 seconds or so. Superimposed on this pattern may be locally generated wind-driven waves arriving at perhaps 5 second intervals.

We have now reached the stage of considering what happens when waves reach the shore. As the depth decreases, the orbital motion of the water below the surface is impeded. The wave is said to 'feel the bottom' and begins to increase in height and its wave length shortens (*Fig 70*). Since the orbital motion is reduced, the front of the wave is 'starved' and becomes steepened and eventually hollowed out so that the crest plunges downward and forward. The falling mass of water traps pockets of air which are compressed, and which then expand with explosive force breaking up some of the water into tiny foam particles. Once the wave has broken, its energy is transmitted into a forward movement of water—a **wave of translation**—and water is thrown up the beach as the

Fig 70 Changes occurring as wave enters shallow water

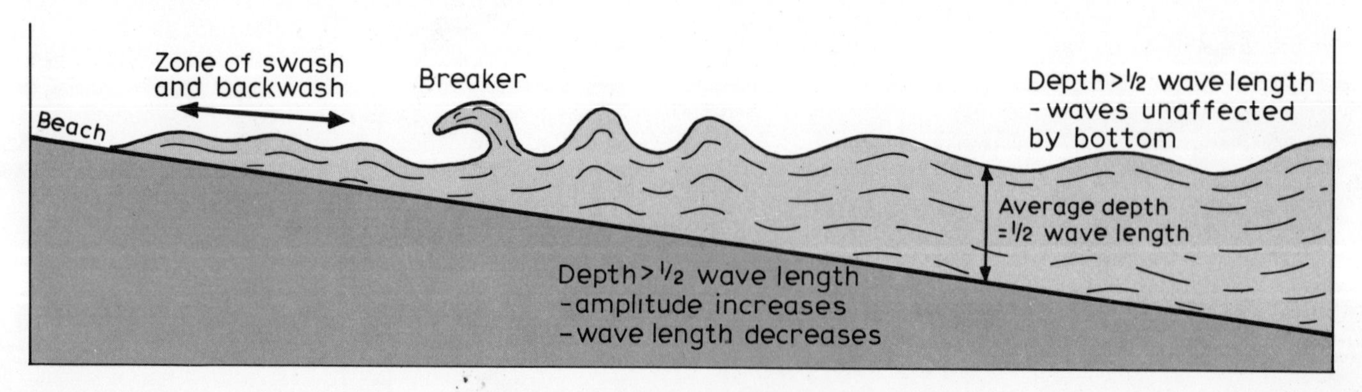

swash. This movement in turn is checked by friction with the beach until all its energy is dissipated. Gravity then draws the water seaward down the beach as the **backwash.** The backwash is likely to be over the surface of a sandy beach, but in the case of a shingle beach the water percolates easily downward and is drawn back to the sea beneath the surface. (*Fig 71*)

From the point of view of the geomorphologist, waves on a beach can be broadly grouped into those which are constructive in their effect, that is they increase the total amount of beach material, and those which are destructive. When **constructive waves** predominate the swash or forward motion throwing material up the beach is more effective than the backwash which draws sand or shingle seaward. Such conditions occur when there is a fairly long interval between the arrival of waves—i.e. there is a long periodicity—of the order of 10 seconds. This allows the backwash of one wave to drain away before the arrival of the next wave, so each swash is thrown forward uninterruptedly. However when waves, such as locally generated storm waves, arrive at close intervals, say 4 or 5 seconds, the swash is checked by the preceding backwash. In addition such waves are frequently of greater height and so plunge downwards more powerfully on breaking. The net effect is that the strong backwash is in almost continuous contact with the beach scouring material downwards towards the sea, hence the term **destructive waves.**

The steepness of the slope of a beach towards the sea also affects the relative importance of swash and backwash, since it will be clear that on a steeply sloping beach the swash will be more quickly checked and the backwash stronger than if the beach gradient is gentle.

It was noted above that wind driven waves with short periodicity are likely to be of the destructive type. This is particularly so in the case of on-shore winds, for then the water is driven higher against the coast with a consequent bottom current away from the shore which assists the removal of material stirred by the waves.

With the alternation of ebbing and flowing tides, changes in wind direction and strength, and the varying impact of long distance swell, conditions on any beach are continually changing. These changes are apparent in almost daily alteration of the form and nature of a beach, and these effects will be considered in greater detail in *Study B3*.

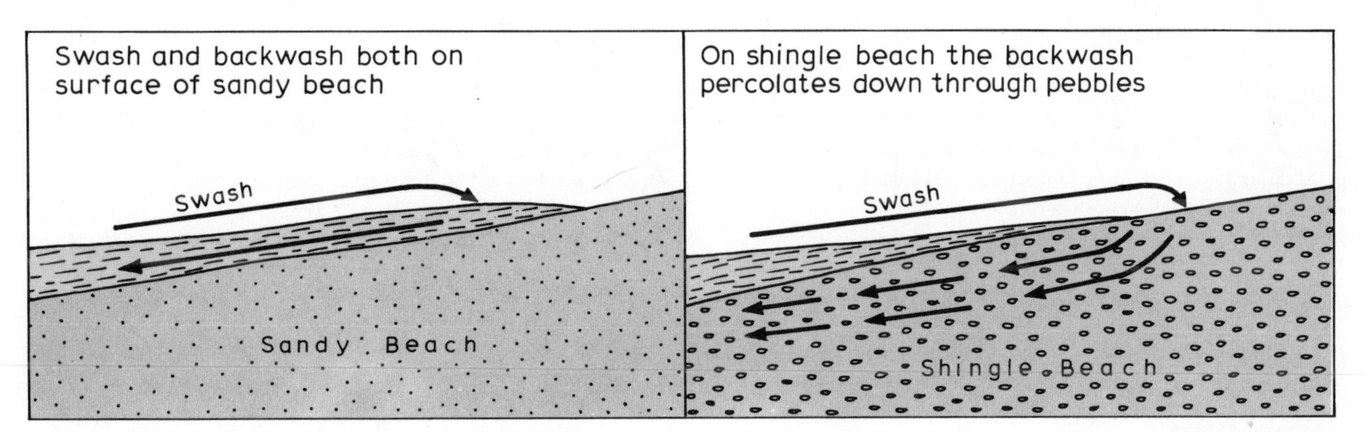

Fig 71 Behaviour of backwash on sand and shingle beaches

FURTHER STUDY

Practical

(1) If it is possible to visit a beach regularly, the student should keep a beach diary, noting such facts as wind speed and direction, the height and periodicity of waves, their direction of approach, e.g. parallel to beach, oblique from SW, etc.; and make plans and sketches of beach features, taking careful note of changes which occur in the shape of the beach or the nature of beach material.

(2) If only occasional visits to the coast are possible, e.g. on holiday, it is still valuable to observe breaking waves closely and to try to determine whether the waves are of constructive or destructive type.

(3) Observation of any stretch of water under conditions of strong wind will provide evidence of wavelet formation and the effect of long and short fetch.

Reading

Dury: *The Face of the Earth*, Ch. 9, pp. 99–102, pp. 106–107
Gresswell: *Beaches and Coastlines*, Ch. 2, pp. 15–33
——: *Physical Geography*, Ch. 25, pp. 327–339
Holmes: *Principles of Physical Geology*, Ch. XXIII, pp. 789–798
Horrocks: *Physical Geography and Climatology*, Ch. 9, pp. 137–139
Monkhouse: *Principles of Physical Geography*, Ch. 10, pp. 232–233
*Sparks: *Geomorphology*, Ch. 8, pp. 168–173
Steers: *Sea Coast*, Ch. 2, pp. 10–22
*——: *Coastline of England and Wales*, Ch. 3, pp. 44–70
Wooldridge & Morgan: *An Outline of Geomorphology*, Ch. XXI, pp. 295–298, pp. 300–301

Freshwater Bay, Isle of Wight

SEA CLIFFS AND STACKS

O.S. 1:63 360 (1.6 cm to 1 km *or* 1 in to 1 mile)
Sheet 180, Solent
O.S. 1:25 000 (4 cm to 1 km *or* 2½ in to 1 mile)
Sheet SZ/38

At the E side of Freshwater Bay on the Isle of Wight (G.R. 350855) the shingle beach of the bay gives way to almost vertical chalk cliffs. The chalk at this point has been subjected to considerable stresses in the geological past and can be seen to dip steeply inland. The cliff at this point is some 30 m in height and is approached from the bay, at low tide, across seaweed strewn rock pools. At close quarters this surface of chalk is not the smooth feature which its name **wave cut platform** suggests. Minor variations in rock resistance, the presence of hard bands of flint within the chalk and the wet piles of seaweed combine to make one pay very careful attention to one's footsteps. But the minor irregularities merge, when viewed from a distance, and the essentially even surface can be seen stretching with a very slight gradient seawards from the foot of the cliffs to the water's edge, with occasional small patches of shingle in minor coves at the foot of the cliff.

Upon this platform the drawing of the area at low tide (*Fig 72*) shows two isolated masses of chalk, each almost as high as the main cliff. One of these **stacks** is in the form of an arch. Close inspection of the cliff face and the stacks reveals a darker zone in the chalk extending from the base of the cliff to a height of some 3 m. The rock in this zone is smoothed and water-worn and clearly marks the rock which will be covered at high tide. A profile of the cliff shows that this zone is slightly undercut compared with the cliff face above and caves are also found in this portion of the cliff. One of these is over 2 m in height, so that a man can walk upright into it, entering at one mouth and emerging from another 25 m further along.

From the form and situation of the features described above it must be clear that all are directly or indirectly the result of erosion by waves. The main features of wave movement have been described in the preceding Study, and the sight of storm waves pounding against a cliff or promenade is familiar to all who live along our coasts and even summer visitors may gain some hint of the enormous power of the waves. Waves erode cliffs against which they break in two main ways. Firstly a breaking wave hurls pebbles and rock fragments against the cliff with great force, and it is this **corrasive** action which is largely responsible for the smoothing effect noticed on rock surfaces between the high and low tide levels. Secondly a wave represents a vast store of energy moving steadily forward. When the wave is checked by a near vertical obstruction such as a cliff this energy is transformed to produce **hydraulic (liquid) pressure,** which in the case of ocean storm waves may reach 30 tonnes per square metre. All rocks contain small air pockets and as the wave rises and breaks, the trapped air is compressed by the hydraulic pressure (*Fig 73*). As the wave falls this pressure is released with explosive force, and jets of spray

Fig 72 Cliffs and wave-cut platform E of Freshwater, I.O.W.

Wave crest approaching cliff. N.B. Some cracks exaggerated for greater clarity

Crest reaches cliff—air pockets trapped in cracks and compressed. Pressure transferred to surrounding rocks

Wave rebounds from cliff—compressed air escapes with explosive force. Cracks enlarged

Fig 73 Action of waves on a cliff face

Fig 74 Undercut cliffs, Portland Bill

Fig 75 Furzy Cliff, N of Weymouth

may be observed from a cliff between the impact of powerful waves. The stresses produced in the rock by such compression and explosive release weaken the adhesion of particles until fragments of rock are prised out from the cliff face to be added to the weapons by which the corrasive process erodes further. Both hydraulic action and corrasion are only significant processes between the levels of high and low tide with a small corrasive effect above high water mark caused by pebbles hurled up by breaking waves, and the main erosive work of the sea can thus be likened to a broad horizontal saw cutting into the land. This is evidenced by the undercutting noted on Freshwater cliffs, and also by the photograph of cliffs at Portland Bill (*Fig 74*). Solution of rocks, especially Chalk and Limestones, by sea water is also a significant process in coastal denudation.

Above high tide mark the cliff is subjected to the processes of **fluvial**[1] erosion and the detailed profile will depend on the relative resistance of the rock, its structural arrangement and its coherence.

Wave erosion at the foot of a cliff, fluvial processes of weathering and erosion, water percolation, the nature of the rock of which the cliff is formed, the dip and jointing characteristics of the rock all combine to produce a wide variety of cliff profiles. As a contrast to the near vertical chalk features described above, we may consider the low cliffs formed in Oxford Clay 5 km N of Weymouth (*Fig 75*). Cliff recession here averages 1 m per year but the main work of the sea in this case is to remove debris produced primarily by the effects of ground water. The Oxford Clay, especially when wet, has not sufficient physical coherence to sustain a very steep cliff face for long, so that masses of clay begin to break away and slump towards the sea—a **rotational slip.** As soon as a fracture zone forms (*Fig 76b*) rain water is concentrated in the crack thus further lubricating the whole mass and encouraging further slumping. Frequently a large portion of the clay becomes so saturated that it flows down the cliff face as a mud glacier. The sea at high water easily fragments the soft debris and removes it, but even if all marine action were to cease, weathering and stream erosion of the cliff would continue for some time until the slope was sufficiently gentle to be sustained by this clay rock.

As any cliff face recedes, the foundations remain, subject now only to relatively slow erosion thus forming the wave-cut platform. As the breadth of this platform increases it will lead waves to break before impact with the cliffs except at times of extreme high tide and storms. So marine erosion of the cliffs is checked and fluvial processes reduce the steepness of the cliff until the slow lowering of the wave-cut platform allows waves to erode once more at the foot of the cliff (*Fig 77*).

Within any cliff forming rock, zones of weakness will exist due to the occurrence of bedding planes, joints, faults, etc. These are more rapidly eroded by the sea to form caves, within which air pressure changes will occur similar to those described above when a breaking wave traps air in crevices in the cliff face. Such caves, progressively deepened by this process and the swirl of tidal waters, may achieve a landward outlet at the top of a cliff—a **blow-hole**—and eventually collapse to form a narrow inlet sometimes referred to as a **geo.** Alternatively, as described at Freshwater, two caves may join within

[1] The terms fluvial-, sub-aerial-, and normal-erosion are variously used to describe the processes of rock destruction, removal and redeposition which are characteristic on land in humid, temperate climates. Since these processes are so closely associated with flowing water (*as shown in Section A*) it is felt that 'fluvial' is the most apt term.

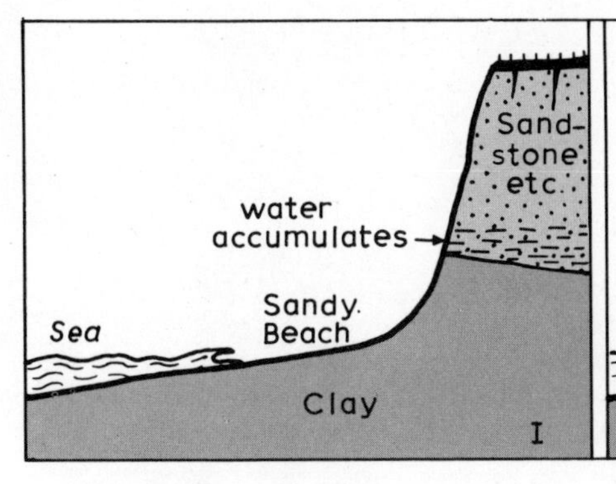

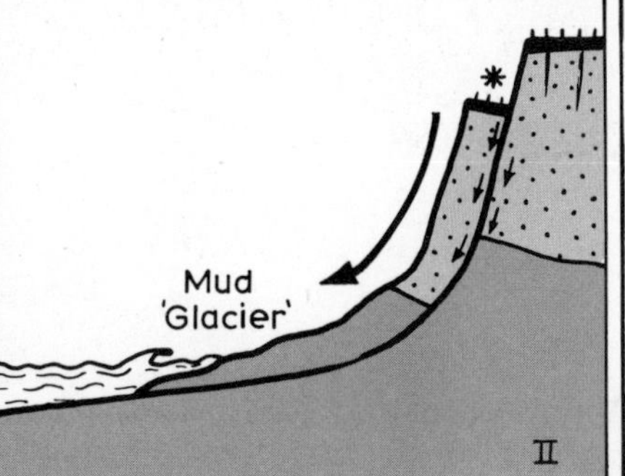

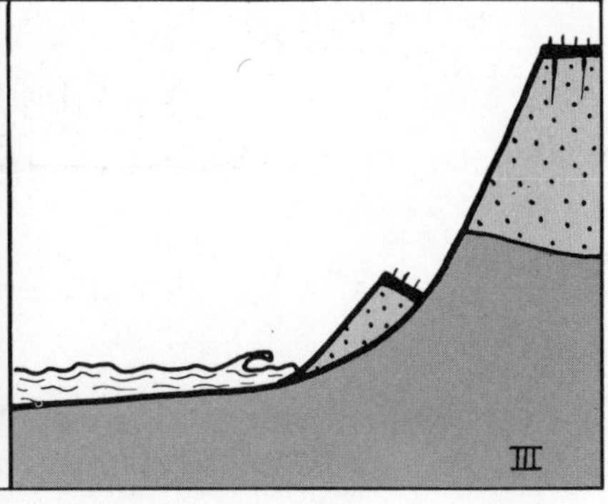

I Rainwater seeps through permeable rock. Underlying clay weakened. Cracks form at surface

II Clay too wet and soft to support steep cliff. Slumping occurs. Water trapped at * helps to lubricate further slumping (rotational slipping)

III Waves remove easily eroded debris. New cracks appear and process is repeated

Fig 76 Cliff erosion by rotational slipping

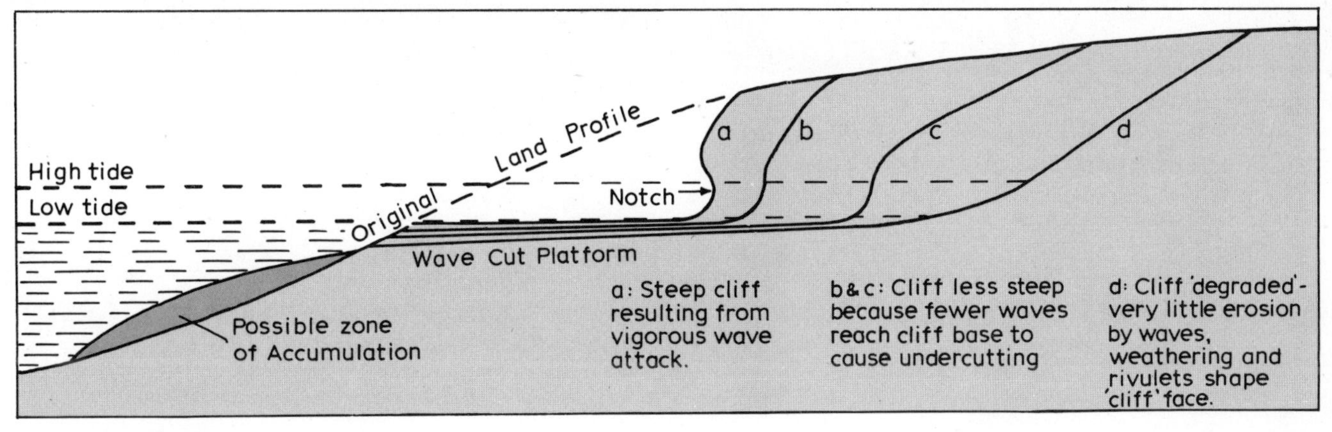

Fig 77 Cliff formation and modification

the cliff or the cave may be eroded right through a headland forming a **natural arch**. The subsequent merging of two geos or the collapse of a natural arch will result in the isolation of a portion of the cliff upon the wave-cut platform. This is a stack. The archway within a stack observed at Freshwater has therefore been formed by the separation of a mass of cliff within which a natural arch was in process of formation and it may be expected that the crest of the arch will eventually collapse, leaving two separate stacks. [1]

[1] In March 1969 a major cliff fall occurred at Freshwater with the result that a third stack, the Mermaid Rock, has been added to the existing Arch and Stag Rocks.

FURTHER WORK

Practical

(1) Cliff features such as those described and explained above may be studied at a great many locations round the British coasts. In few cases can details of cliff formation be interpreted accurately from Ordnance Survey maps, although stacks and wave cut platforms can often be recognized on a scale of 1:25 000. Any available photographs of cliff coasts should be studied, sketches made to show significant features and explanations of these attempted.

(2) If coastal cliffs can be visited, make a careful record of their height; the angle of the cliff face (use a clinometer); any evidence of undercutting, slumping, etc.; the nature of the rock involved (consult a Geological Map). Try to relate evidence of coast erosion to exposure to wave attack.

(3) If possible view the film and/or filmstrip *Against the Sea* (Rank), and consider the problems presented.

Reading

Dury: *The Face of the Earth*, Ch. 9, pp. 102–104
Gresswell: *Beaches and Coastlines*, Ch. 3, pp. 34–51
——: *Physical Geography*, Ch. 27, pp. 365–375
Hardy & Monkhouse: *Physical Landscape in Pictures*, pp. 70–74
Holmes: *Principles of Physical Geology*, Ch. XXIII, pp. 798–805
Monkhouse: *Principles of Physical Geography*, Ch. 10, pp. 233–239
Scovel: *Atlas of Landforms*, pp. 144–145
*Sparks: *Geomorphology*, Ch. 8, pp. 173–184
*Steers: *Sea Coast*, Ch. 5, pp. 61–96
——: *Coastline of England and Wales in Pictures*, Nos. 31, 33, 53, 123, 133, 143, etc.
Strahler: *Physical Geography*, Ch. 27, pp. 412–417
Wooldridge & Morgan: *An Outline of Geomorphology*, Ch. XXI, pp. 301–303

Fig 78

BEACH PROCESSES

Study B3

Sit on any beach on a summer day and watch the effect of the breaking waves. As the swash washes up the beach the acceleration of movement in the water disturbs particles of sand or shingle, lifts them temporarily into suspension only to be deposited again as the water falls back across the surface of a sandy beach, or percolates downwards through shingle (*Fig 71*). Hardly an unfamiliar or dramatic scene; not to be compared with an eruption of Mount Etna or the thunderous roar of Niagara Falls. Nevertheless some simple mathematics reveals the significance of the process we have observed. If the very moderate assumption is made that one wave disturbs 1 kg of beach material for each metre of its length, and that six waves are breaking every minute, we find that in 24 hours the total amount of material moved for each kilometre of coastline is 8640 tonnes! What then might the result be if one considered storm waves crashing against a beach? This gives some idea of the potential movement which can occur on a beach, even if it is true that a proportion of the material disturbed is redeposited in the same position as that from which it was removed.

To return to our observation of breaking waves on the beach: it is often possible to follow the movement of a small pebble on a sandy beach or a distinctive stone on a shingle beach, and to establish its overall direction of movement. This may be either up or down the beach, or sideways along the beach. These two forms of motion will now be considered separately, though in fact both may occur at the same time.

As has been described above (*Study B1*) the nature of the waves and their periodicity largely determines whether they are constructive or destructive in character —that is whether they add to or remove beach material. On shingle beaches which normally slope seawards at a steeper angle than sandy beaches, the changing nature of wave action can be observed by studying the profile or cross-section of the beach. This typically shows a number of elements (*Fig 79*). The inland limit of the beach is in the form of a ridge composed of pebbles coarser in size than those found nearer to the sea. This **storm ridge** or **berm** represents material pushed up the beach by constructive waves at the highest tides. It may reach to well above Spring High Tide mark because it can be added to under storm conditions by shingle thrown with spray high above the level of the actual waves. Thus Chesil Beach crest (*Study B5*) reaches a height of 13 m above

Fig 79 Characteristic profile of shingle beach

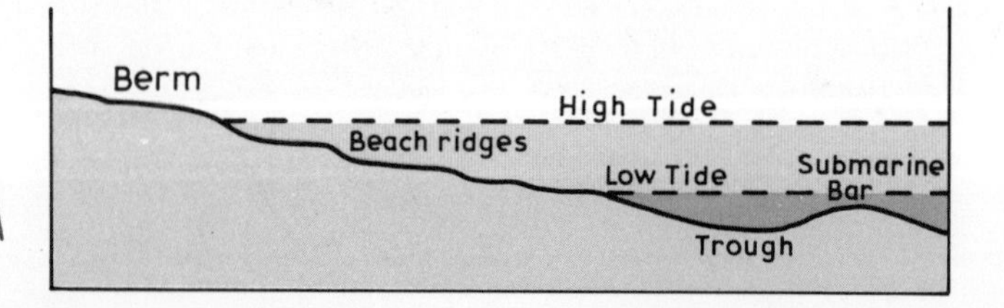

high water. The seaward side of the berm is often marked by a sharp increase of gradient produced by the removal of material by destructive waves. This alternation of ridges, resulting from a period of construction, and steep seaward slopes indicating the later removal of beach material by storm waves is often repeated a number of times across the foreshore; with each ridge representing the limit reached by constructive waves under the tidal and wind conditions then obtaining. Such ridges tend to be composed of slightly larger pebbles than are found to the seaward side of them since it is these large pebbles which are most likely to be thrown forward by the swash and then left stranded, too large to be dragged back by the weaker backwash of constructive waves. Below low tide level, in the offshore zone, one or more submarine ridges, often composed of much smaller particles, may be formed as a result of the disturbance of sediment by breaking waves.

It will be clear that a beach profile is constantly changing to adapt to varying conditions of tide, wave frequency, wave height, etc. Berm ridges created by long periodicity constructive waves in summer may be destroyed by winter storms; the character of a favourite holiday beach may change markedly from one summer to the next. Long-term observations can indicate that some beaches are, on average, gaining or losing material. In this context it may be noted that it took at least six years for Lincolnshire beaches destroyed in the Tidal Surge of 1953 to regain their former condition.

Seaward of the line at which waves break, currents may be sufficiently strong to carry significant amounts of debris seaward or shoreward. The most important of these are the **rip currents** which flow strongly seaward through gaps in a submarine ridge, and which may be powerful enough to be a real danger to swimmers. (*Fig 80*).

When studying a beach profile one may notice that the storm ridges do not always have a seaward face entirely parallel to the shore, but that this slope is indented to form minor 'bays and promontories' (*Fig 90*). These scallops are known as **beach cusps**, and may attain a relief of 3 m and extend for 18 m between the 'headlands'. It is easy to see how such features, once initiated, may persist for some period, since each wave washes most strongly into the hollow, carrying larger pebbles to either side and leaving the depression floored by finer particles. The author has observed that below the water line the pattern of cusps is often marked by a complementary pattern of delta like mounds, which, taken in conjunction with the sharp points of the headlands, suggests that cusp formation is part of a destructive process by which material is being scooped downward from the beach. How beach cusps originate is not altogether clear. They seem to occur irrespective of whether waves are advancing parallel to the beach or at an angle to it, and in widely varying conditions of wave type. Perhaps a clue may be found in the scalloped edge of the swash even on a flat, sandy beach. Tentatively one wonders whether the interaction of two or more sets of waves from differing directions may provide a clue to their origin.

Fig 80 Schematic indication of the relation of rip currents, R, to depressions in the submarine offshore bars, B

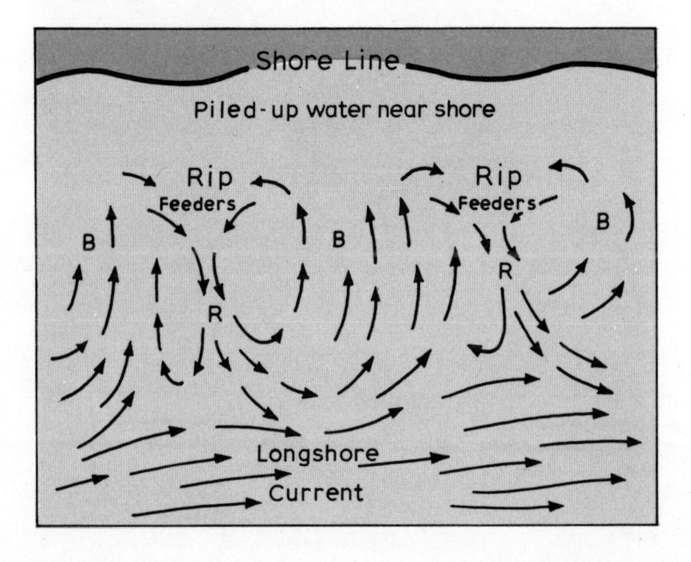

In addition to the movements of beach material up and down the foreshore described above, an observer may notice that pebbles are shifted longitudinally along the coastline. Individual pebbles may be seen to be moved beachwards and sideways by each swash and to be drawn straight back by the succeeding backwash so that they come to rest some distance to one side of their original position. The effect of each wave may be small, say a movement of 10 cm, but with ten waves per minute this could total almost 1.5 km per day, and we have already noted the great volume of material which may be so moved. This lateral movement, or **longshore drift**, will only occur when the line of breaking waves is not parallel to the shore. The speed of wave advance is reduced as the wave enters shallow water because of friction between the orbiting particles and the sea bed. Thus waves approaching a gently shelving shoreline at a sharp angle become slewed round and may be almost parallel to the beach by the time they break (*Fig 81*), hence the way in which waves mould themselves to the shape of a bay. Such a change in direction is termed **wave refraction** (*Fig 83*). But parallelism is not always completely achieved, in which case longshore drift is liable to occur.

Evidence of longshore movement of beach material is not hard to find—especially when some natural or artificial obstacle checks the movement. In an attempt to prevent loss of beach material which may be a tourist attraction or a protection against wave erosion many coastal local authorities have erected groynes—barriers of wood, steel or stone running from the sea-wall across the beach. Longshore drift piles sand and/or shingle against one side of these groynes; in some cases there may be a difference in level of 2 m between the beach on either side of a groyne. The photograph (*Fig 82*) shows this clearly in the case of Brighton with a predominant longshore drift from W to E under the influence of waves approaching from the SW. It may be noted that

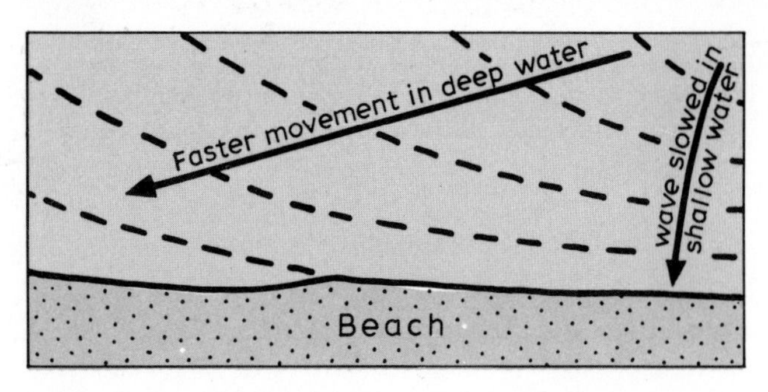

Fig 81 Adjustment of wave crests towards parallelism with beach

two stormy days with a SE wind could completely alter this pattern, shifting thousands of tonnes of beach material and piling them on the other side of the groynes. A further point to notice is the tendency of the line of the water's edge to swing round to approximately right angles to the direction from which the dominant waves are approaching. Groynes are discussed here because they provide easily observed evidence of the occurrence of longshore drift—headlands, river mouths, harbour breakwaters, etc. have a very similar effect.

Around the British Isles it is generally true to say that longshore drift is eastward along the S coast, northward along the W coast, and southward along the E coast, but there are many local variations resulting from the shape of the coastline, e.g. along the N Norfolk coast the predominant movement is westward towards the Wash. The influence of longshore drift is further considered in *Studies B4, B5* and *B6* in its effects on spits and other coastal forms.

Further contributory factors in the movement of sand, mud and shingle along the shore are **offshore currents.** Around our shores these are often associated with tidal movements which are often more effective in one direction than another due to irregularities in the coastline. Currents consequent upon the outflowing of river water and other local factors may produce similar effects and the net result is a highly complex pattern of water movements. Such currents do not often have the competence to move large shingle, and most large coastal depositional forms appear to owe much more to longshore drift than to offshore currents.

Fig 82 Longshore drift at Brighton

Fig 83 Refraction of waves within Lulworth Cove, Dorset

FURTHER WORK

Practical

(1) Study O.S. 1:63 360, Sheet 116 (Dolgellau). Determine the lines of low water mark of major beach formations and check the relation of these lines to the direction of longshore drift, and to the direction of maximum fetch. This exercise can be carried out also on the 1:250 000 map of the area, and may be repeated for other areas.

(2) If access to a beach is possible, make accurate measurements to determine the beach profile. Relate this to the nature of beach material. By careful observation discover whether the beach is being built up or eroded at the time of your visit, and whether there is evidence of longshore drift. Is any cusping evident? It may be possible to obtain evidence of longshore and 'up and down' movements of beach material by tracing the dispersion of a number of painted pebbles placed at a point at the water's edge. (Research workers use this basic method with the more sophisticated refinement of tracing radioactive particles with a geiger counter!) Note the suggestion following *Study B1* for keeping a Beach Diary.

Reading

Gresswell: *Beaches and Coastlines*, Ch. 5, pp. 71–79
——: *Geology for Geographers*, Ch. 25, pp. 331–337
——: *Physical Geography*, Ch. 26, pp. 340–343
Hardy & Monkhouse: *Physical Landscape in Pictures*, pp. 74–75
Holmes: *Principles of Physical Geology*, Ch. XXIII, pp. 808–820
Monkhouse: *Principles of Physical Geography*, Ch. 10, pp. 239–244
Scovel: *Atlas of Landforms*, pp. 130–131
Sparks: *Geomorphology*, Ch. 8, pp. 184–188
Steers: *Sea Coast*, Ch. 2, pp. 10–22
*——: *Coastline of England and Wales*, Ch. 3, pp. 44–63

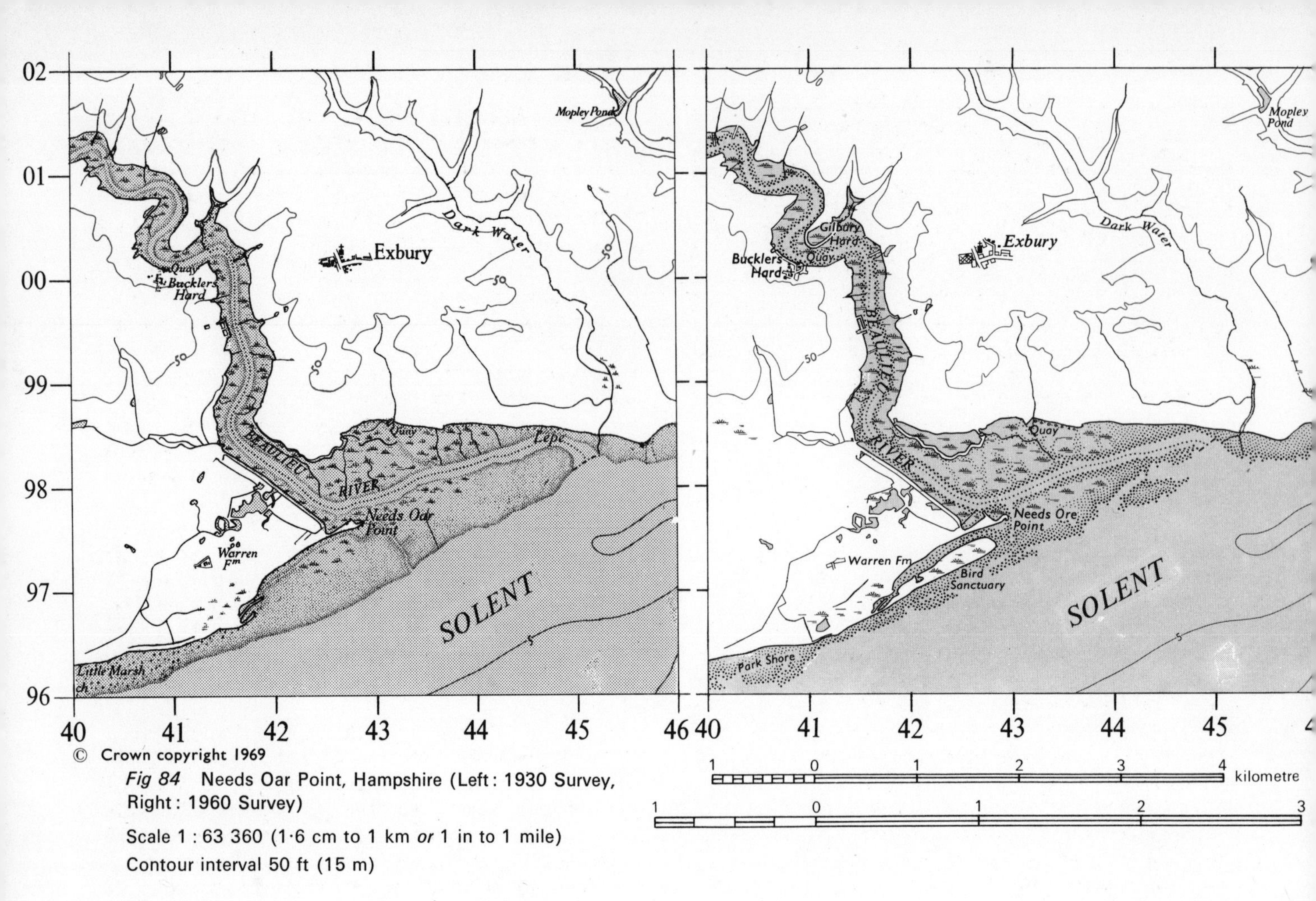

© Crown copyright 1969

Fig 84 Needs Oar Point, Hampshire (Left: 1930 Survey, Right: 1960 Survey)

Scale 1 : 63 360 (1·6 cm to 1 km *or* 1 in to 1 mile)
Contour interval 50 ft (15 m)

Needs Oar Point and Hurst Castle Spit, Hampshire

Study B4

SHINGLE FORMATIONS; BARS AND SPITS

O.S. 1 : 63 360 (1.6 cm to 1 km *or* 1 in to 1 mile) Sheet 180
O.S. 1 : 25 000 (4 cm to 1 km *or* 2½ in to 1 mile) Sheets SZ/49, Needs Oar Point, SZ/39 and SZ/38, Hurst Castle Spit

This study is based upon shingle features resulting from marine transport and deposition on the northern shore of the Solent, and leads to a discussion of other shingle formations.

The above maps show Needs Oar Point which lies at the mouth of the Beaulieu River as it is recorded on two editions of the 1 : 63 360 O.S. map. It will be evident that a marked change has taken place in the thirty years separating the revisions of this sheet—a pointer to the fact that such features can develop, or diminish, quite rapidly. In many studies of coastal features of deposition, valuable evidence of change exists on old maps—in some cases seventeenth- and eighteenth-century maps can provide data, though allowance must be made for varying standards of accuracy.

Needs Oar Point as shown in 1930 is composed of fine shingle and sand, and has remained almost unaltered in appearance. But in the last two decades a shingle spit more than 1 km in length at high tide has been built out parallel to the old shoreline on its seaward side and on the spot investigation suggests that this feature is still growing actively. In the field it is noticeable that there is a series of low ridges and hollows running diagonally across the landward side of this new spit (*Fig 85*). These shingle ridges are no more than 0.5 m above the intervening hollows, but vegetation in the form of grasses is beginning to gain a foothold in the hollows. By analogy with similar features elsewhere it is reasonable to suggest that each ridge marks one stage in the growth of the spit eastward—in other words each ridge was once the end of the spit until supplanted by later deposition.

The growth of both the original Needs Oar Point and its more recent parallel development has been from the W bank of the Beaulieu River, thus diverting the river's outlet to the E. The shape of a spit at a river mouth must be the result of the interplay of current and wave action in the sea and the flow of river water. In seeking an explanation for the eastward growth of this spit one must consider the position of Needs Oar Point in relation to surrounding features. To the S there are only 3 km of

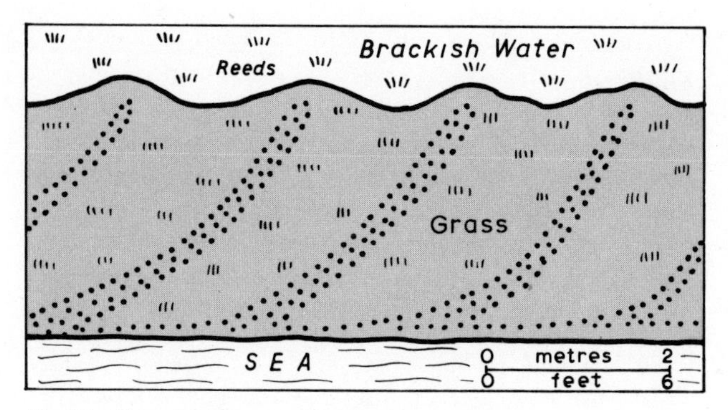

Fig 85 Shingle ridges on Needs Oar Point

water before the Isle of Wight is reached; to the E there are 20 km of water across the Solent to Portsmouth, whereas to the SW the Solent lies open to 6000 km of the Atlantic Ocean. Add to this the predominance of SW winds and it is understandable that the dominant waves will be those approaching from the SW. Such waves do not become completely adjusted to the shape of the coastline and are thus likely to produce some W–E drift of material along the coast. It must not be thought that such longshore drift is the only process of significance. If this were the case, how can one explain the relatively sudden development of the new spit? And where is the source of the pebbles of which it is formed? However the fact that many shingle spits of this type along the S coast of England have grown from W to E, and the evidence of beach drift where groynes, etc. have been built, do suggest that longshore drift related to the direction of dominant waves is an important factor to be considered.

A feature with some characteristics in common with Needs Oar Point can be seen at the mouth of the R. Adur at Shoreham, Sussex (O.S. 1:63 360, Sheet 182 and *Fig 86*). Here too a shingle spit has extended eastward across the mouth of a river, but in this case the spit reached a maximum length of 6 km until the river outlet was pushed as far E as the W end of Hove. Then in mediaeval times documentary evidence records the breaching of the spit by high storm waves at the present mouth of Shoreham Harbour. The E arm of the harbour partially silted up until it was cleared by dredging in recent times.

A second type of spit formation is that in which a sand or shingle spit grows out at a marked angle to the shore line. Hurst Castle Spit at the W end of the Solent (*Fig 87*) extends out from a point where the coast changes direction from WNW–ESE to SW–NE, and consists of a narrow shingle ridge extending NW–SE for about 2 km before broadening out and curving back on itself. On the SW side and at the tip of the spit there is a steep offshore slope to fairly deep water; on the NE side, within the curve of the spit the water is shallow and a considerable expanse of mud is exposed at low tide. These salt marshes or **saltings**, are partially colonized by plants which themselves serve to trap more silt and thus encourage the build-up of mud. The whole area is intersected by a complex network of slightly deeper channels some of which retain water at low tide.

The differences between the two sides immediately suggests that different processes are dominant on the seaward and the Solent sides. As in the case of Needs Oar Point it seems most likely that eastward longshore drift along the coast is a major process, and that the growth outward from the land is related to the fact that waves cannot immediately adjust to a sharp change in the angle of the coast.

From about 2 km out from the land, a series of shingle ridges run back from the main beach ridge in a NNW direction, generally increasing in length until the ridge forming the present limit of the spit is reached. The main spit is aligned approximately at right angles to the direction of maximum fetch—from the SW—and using the same argument the hooks or **recurves** at the end of the spit can be said to be aligned at right angles to the second longest fetch—from the NE down the Solent. Each barb on the hook is likely to represent a past limit of the spit. The very short fetch to the Isle of Wight to the SE means that waves from this direction have little effect—and so there is no straight stretch of beach facing in this direction—hence the sharp angle at the tip of the spit.

The processes operating to produce shingle features such as those described above are not yet fully understood, though it is clear that the present form of such features represents a state of approximate balance between wave action (longshore drift, length of fetch, constructive and destructive waves), offshore currents such as those produced by rivers or tidal flow, the supply of beach material produced by coastal erosion either locally or from some distance away, and changes in sea-level in geologically recent times. Even these factors may not represent all the influences operating on a coastline; hence the danger of attributing observed features to the operation of any one process.

Fig 86 Changes in the mouth of the R. Adur, Sussex

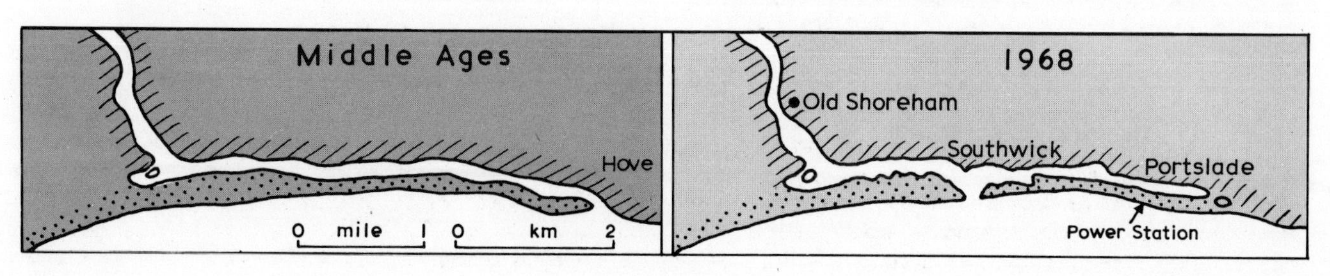

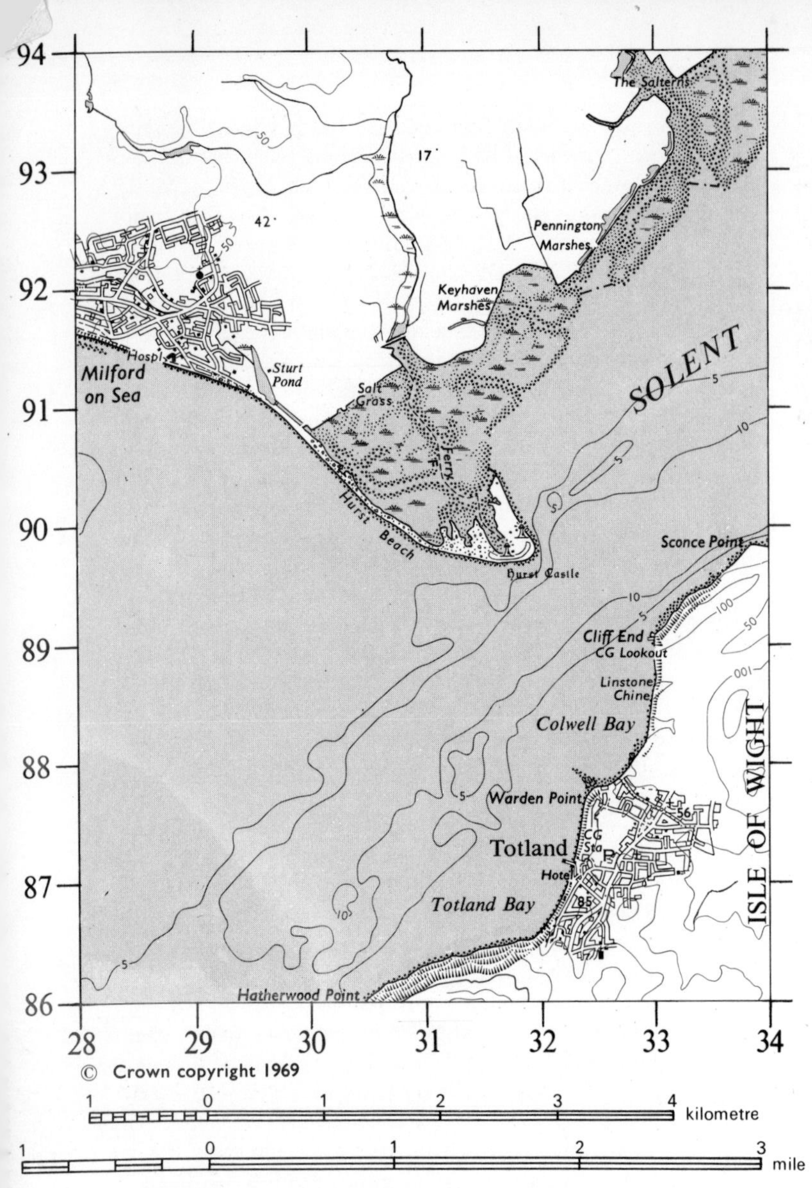

Fig 87 Hurst Castle Spit, Hampshire
Scale 1 : 63 360 (1·6 cm to 1 km *or* 1 in to 1 mile)
Contour interval 50 ft (15 m)

Dungeness Foreland (O.S. 1 : 63 360, Sheet 184) was at one time attributed to the fact that shingle accumulated where the eastward drift up Channel met the southward drift characteristic of the E coast. It can, however, be easily shown that eastward drift continues from Dungeness along the coast towards Dover. Study of the shingle ridges of Dungeness (*Fig 88*) indicates erosion of older ridges on the W side of the foreland and the construction of newer ridges on its E coast. Similar evidence suggests that the Foreland now has a much more pointed shape than it had in the past, and this is supported from historical records, including maps dating back to A.D. 750. It appears that at an early stage a spit (C–C_{I}–C_{II} *on Fig 88*) stretched SW–NE across a broad bay now marked by Romney Marsh, with a gap somewhere in the E end through which the R. Rother reached the sea. By Roman times this spit had developed a seaward curve (E–E_{I} *Fig. 88*) and the Rother changed to its present course. The curve of the Foreland has sharpened—perhaps under the influence of two sets of dominant waves (a) up Channel from the SW, (b) through the Straits of Dover from the NE. A further factor is the nearness of the coast of France to the SE, so that waves from this direction are ineffective (cf. the sharp angle at the tip of Hurst Castle Spit). It remains true that we are not able to explain why Dungeness should have begun to grow outward from the coast while other shingle spits have not done so.

Fig 88 Dungeness Foreland. Larger scale map shows alignment of shingle ridges. On both maps letters A-A, B-B etc and dates indicate positions of progressively later shorelines. (After Holmes)

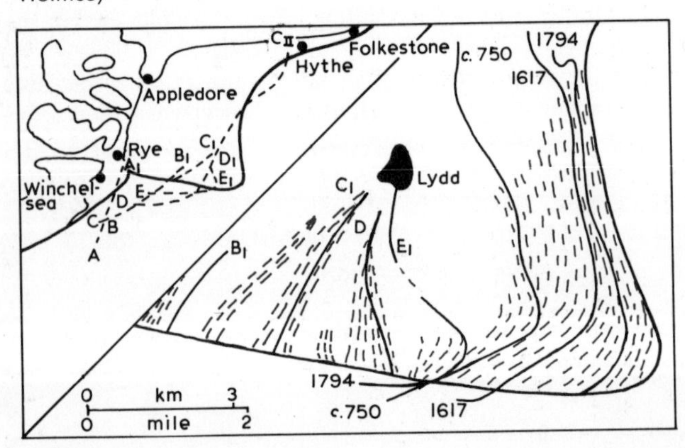

FURTHER WORK

Practical

(1) Use O.S. 1 : 63 360, Sheet 150, to write a description of Orford Ness, illustrated by an annotated sketch map.

(2) Having described this feature, read Steers *Coastline of England and Wales*, pp. 385–389 and then write a short account of the formation of Orford Ness. (See also G. de Boer and A. P. Carr, 'Early maps as historical evidence for coastal changes', in *Geographical Journal*, Vol. 135, March 1969.)

Reading:

Gresswell: *Beaches and Coastlines*, Ch. 5, pp. 79–85
——: *Physical Geography*, Ch. 26, pp. 344–353
Holmes: *Principles of Physical Geology*, Ch. XXIII, pp. 820–824, 826–828
Monkhouse: *Principles of Physical Geography*, Ch. 10, pp. 244–248
——: *Landscape from the Air*, pp. 39, 40.
Scovel: *Atlas of Landforms*, pp. 128–129, 138–141
Sparks: *Geomorphology*, Ch. 8, p. 198–205
Steers: *Coastline of England and Wales*, Ch. 8, pp. 294–295 and 318–331
——: *Coastline of England and Wales in Pictures*, Nos. 15 and 17
——: *Sea Coast*, Ch. 7, pp. 162–166
Strahler: *Physical Geography*, Ch. 27, Ex. 4, p. 432
Wooldridge & Morgan: *An Outline of Geomorphology*, Ch. XXI, pp. 306–317

Chesil Beach, Dorset

Study B5

AN OFFSHORE SHINGLE RIDGE

O.S. 1:63 360 (1.6 cm to 1 km *or* 1 in to 1 mile) Sheet 177, Lyme Regis and Sheet 178, Dorchester

The so-called 'Island' of Portland is approached from Weymouth by a bridge and causeway carrying a road (A354) and a disused railway. But apart from this man-made feature, Portland is also 'tied' to the mainland by the great shingle ridge of Chesil Beach (or Chesil Bank). At the Portland end this ridge rises to about 13 m above O.D. and is composed of large pebbles averaging about 7 cm in diameter. Its cross-section above water level shows a series of ever-changing storm ridges, surmounted by a final uppermost ridge, or berm, followed by a steep backslope dropping towards the road and Portland Harbour. In spite of the height of the shingle, storm waves on occasions break right over it, and when such conditions pile water up on its seaward side, water passes through the shingle to emerge as temporary gullies known as 'canns' on the backslope to such an extent that the road may be closed by flooding.

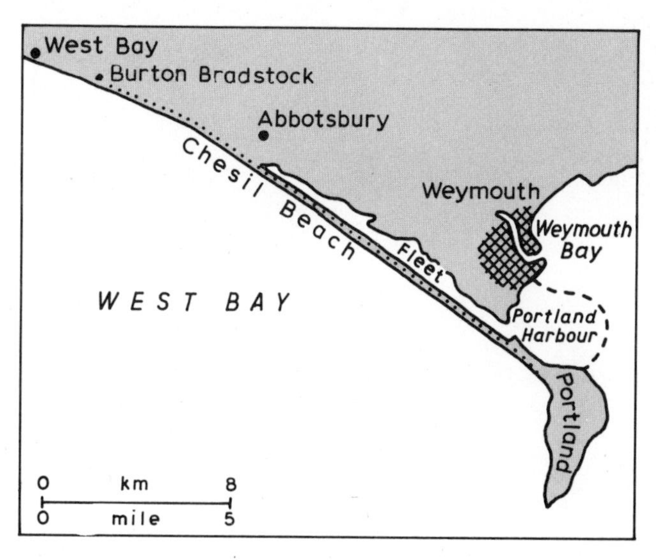

Fig 89 Location of Chesil Beach

Fig 90 Chesil Beach from Portland

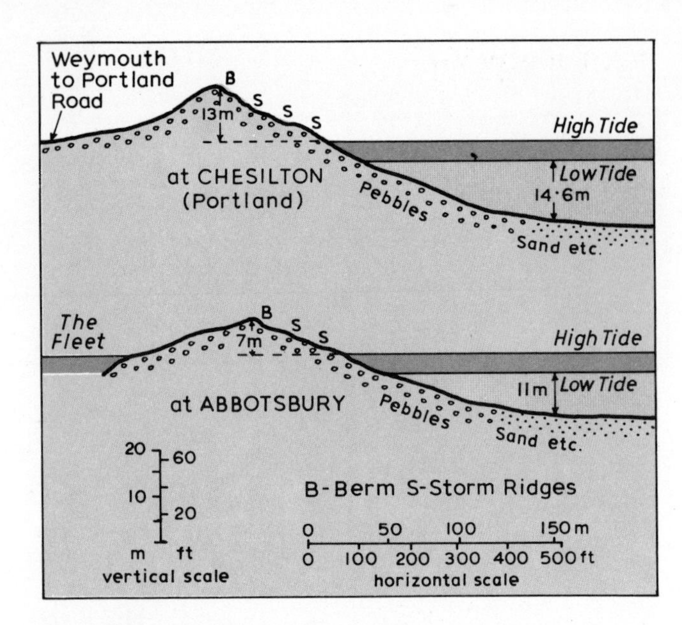

Fig 91 Sections through Chesil Beach

Between the road bridge (667760) and Abbotsbury (568840), Chesil Beach is separated from the mainland by a 'lagoon' of brackish water known as the Fleet, varying between 100 m and 1000 m in width. The Fleet is floored by recent alluvium and peat to considerable depths before firm clay is reached. As one travels NW along Chesil Beach the maximum height gradually decreases to about 7 m above O.D. at Abbotsbury, and the shingle size also decreases to about 2.5–3 cm in diameter.

The decrease in height and in pebble size continues from Abbotsbury to Burton Bradstock. In this section the beach is joined to the land. A miniature cliff of a metre or so in height faces the shingle and this is followed by a remarkably even and undissected slope of 8° or 1 in 6½. At Cliff End (near Burton Bradstock) we reach what is considered to be the end of Chesil Beach proper. From here westwards the coastline is one of high steep cliffs fringed by a narrow beach of very fine shingle. Investigation of early maps suggests that these cliffs have retreated at an average rate of 0.5 m per year over the past century.

The pebbles which form Chesil Beach are of many different kinds, derived from a great variety of sources. These include granites from Devon and Cornwall, pebbles from the Budleigh Salterton area, flint from chalk areas, and chert from the Portland Series, to mention only a few, though it is flint which is the predominant beach material.

Finally, in description of this feature, unique within the British Isles, it is necessary to consider its position in relation to more distant features. The importance of fetch has been referred to above (*p. 57*) and *Fig 92* shows that waves approaching Chesil Beach from the SW between Start Point and Ushant have an uninterrupted fetch of many thousand kilometres across the Atlantic Ocean. It can also be seen that, in accordance with general theories of beach formation, Chesil Beach particularly in its southern portion is aligned at approximately right angles to the direction of maximum fetch. Waves approaching from the SSW have a fetch from the Breton coast of 270 km, and from the Channel Isles due S the fetch is 120 km. Clearly waves from due W across Lyme Bay will be relatively insignificant in their effects.

The processes by which Chesil Beach has been constructed have been a matter for geomorphological debate for many years. Much attention has been paid to the grading in pebble size from large stones near Portland to very fine shingle at Burton Bradstock. This grading is so consistent that local seamen claim to be able to locate their position on the beach by reference only to the size of the shingle. Such a precise grading would appear to suggest a steady longshore drift of material from Portland northwards, but analysis of the pebbles shows that only a tiny fraction can have derived from Portland itself. It may be that, under present conditions, the waves from the SSW and S directions cause northward longshore drifting, but that due to their limited fetch they tend only to move the smaller pebbles. On the other hand the waves along line AB (*Fig 92*) which are much more powerful, will cause a general southerly drift of material of all sizes. The net effect would thus be to grade the shingle in accordance with the observed distribution. The fact that waves along line AB are not quite approaching the Chesil Beach at right angles, particularly in its N portion, suggests that the whole feature has not yet reached a stable state.

This still leaves the problem of the origin of the beach unanswered, one hypothesis suggests that in order to solve this problem one must look back to events in the

Fig 92 Fetch of waves affecting Chesil Beach

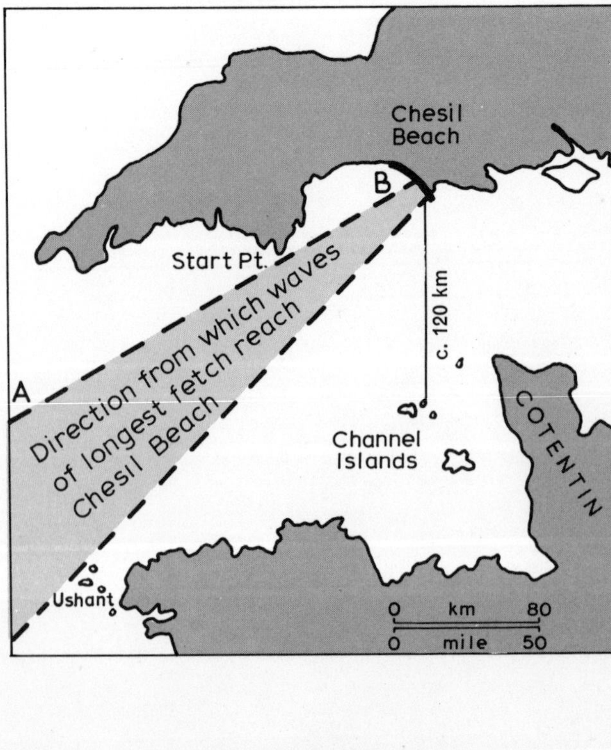

geologically recent past.[1] During the last glacial epoch, 18 000–20 000 years ago, sea-level fell to at least 100 m below its present level, so that the whole of Lyme Bay and indeed much of the English Channel was dry. At this time intense cold and solifluction (*see below p. 111*) produced vast quantities of rock debris from land which now surrounds the bay and this debris, added to by beach material from the pre-Glacial shore line, would have been carried out into the dry floor of the bay by rivers formed during the brief summers, and as the climate slowly ameliorated. Thus a great mass of rock fragments was spread from the surrounding land over the bed of Lyme Bay. As sea level gradually rose at the close of the Ice Age this debris was slowly moved inshore, and under the influence of the waves with the greatest fetch became concentrated in the eastern part of the bay, trapped by the protrusion of Portland, and gradually swung round to its present position under the influence of the dominant SW waves. The rapid recession of cliffs in the Burton Bradstock area at the present time referred to above may be part of a process by which the line of Chesil Beach is gradually slewing round towards a direction more precisely at right angles to the direction of maximum fetch.

In the SE section of the beach shoreward movement of the shingle ridge is insignificant. Even on the fairly rare occasions when shingle is thrown right over the beach in storms the quantities involved are very small—probably less than the amount dragged down the back-slope by holidaymakers scrambling up the beach! Research into bore holes, etc. shows the likelihood that Chesil Beach has now been pushed back to the approximate line of an earlier coast, and in the S at least is now fairly static.

[1] The author acknowledges gratefully the work of G. C. Poole —unpublished Cambridge University dissertation—which has been used in the compilation of this Study.

FURTHER WORK

Practical

(1) In the preceding Study one explanation of the observed features of Chesil Beach has been outlined. Read as many of the references given below as possible and list the differences between the explanations given. Try to suggest observations and experiments which would test the validity of the conflicting theories. (Remember the scientific dictum that it is virtually impossible to prove that an hypothesis is true, but that a single experiment or observation may prove it to be false.)

(2) If a visit to a shingle beach is possible, it is fascinating to attempt to determine the origin of beach pebbles, and to consider what application your results have to the process of beach formation. A useful guide for such a study is *Pebbles on the Beach* by C. Ellis (Faber paperback 1965).

(3) Use atlas maps to test the statement that 'the line of a beach tends to be at right angles to the direction of maximum fetch.'

Reading

Gresswell: *Beaches and Coastlines*, Ch. 5, pp. 85–86
Holmes: *Principles of Physical Geology*, Ch. XXIII, p. 824
Monkhouse: *Principles of Physical Geography*, Ch. 10, p. 247
Sparks: *Geomorphology*, Ch. 8, pp. 201–202
Steers: *Sea Coast*, Ch. 7, pp. 156–162
*——: *Coastline of England and Wales*, Ch. 8, pp. 271–277
——: *Coastline of England and Wales in Pictures*, No. 30
Trueman: *Geology and Scenery* Ch. 8, p. 120.

Woolacombe Bay, North Devon

Study B6

BAYS AND HEADLANDS

O.S. 1 : 63 360 (1.6 cm to 1 km *or* 1 in to 1 mile) Sheet 163, Barnstaple
O.S. 1 : 25 000 (4 cm to 1 km *or* 2½ in to 1 mile) Sheet SS. 44/54

MAP EXERCISE (*Fig 93*).

(1) Draw a map of this section of the coastline. Use a heavy line to show cliffs, and hachures to indicate other steep slopes behind High Tide mark. Use suitable symbols to show the location of (i) sand, (ii) rocks uncovered at low tide. Use the geological sketch (*Fig 94*) to show rock outcrops on your map. Summarize your findings in a short written statement.

(2) In the light of the preceding work, comment on the validity of each of the following statements:

(a) The major headlands of Morte Point and Baggy Point coincide with the outcrops of Morte Slates and Baggy Beds.
(b) Cliffs and offshore rocks are to be found mainly at the headlands.
(c) Sandy beaches occur at the head of bays.

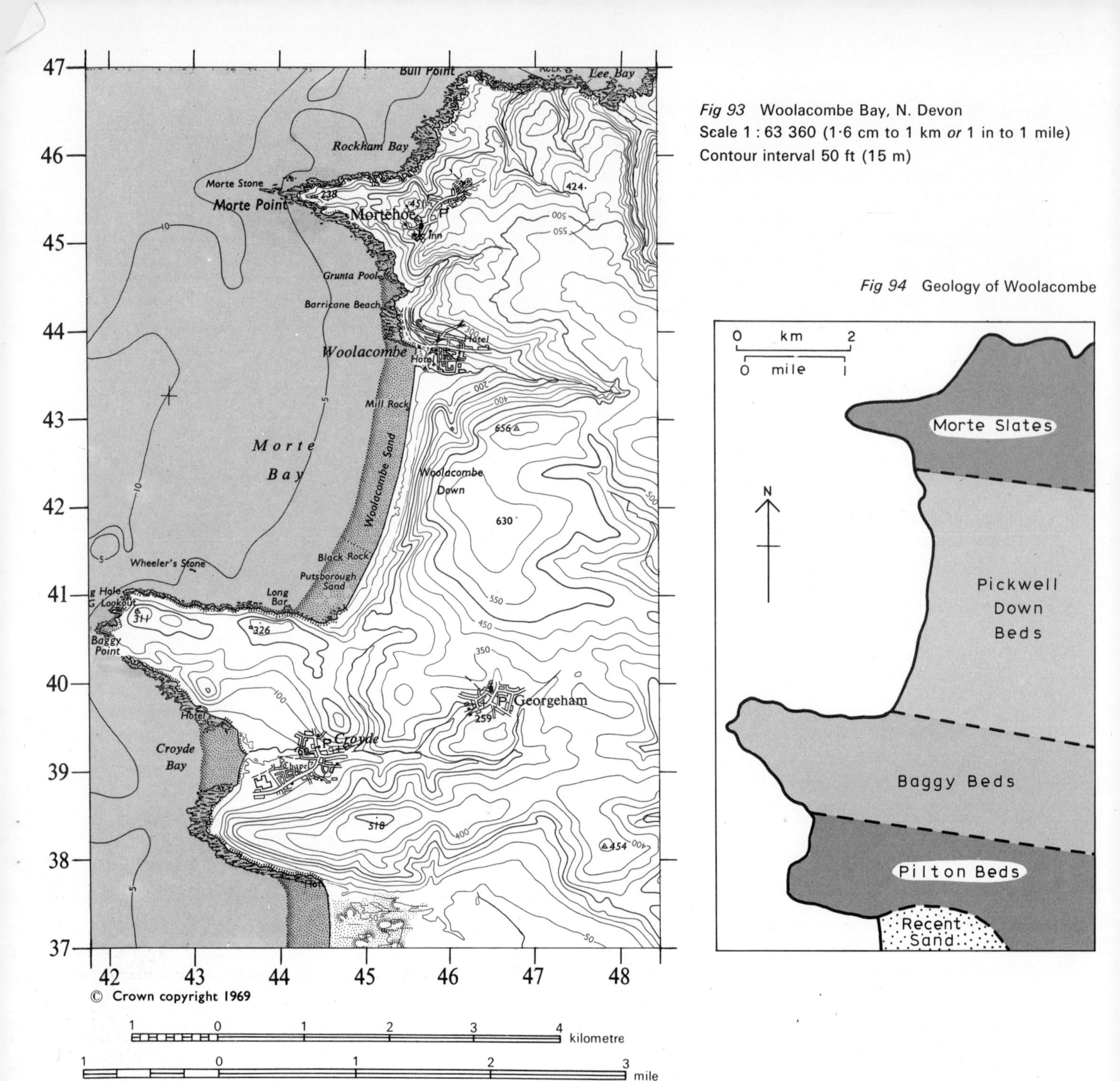

Fig 93 Woolacombe Bay, N. Devon
Scale 1 : 63 360 (1·6 cm to 1 km *or* 1 in to 1 mile)
Contour interval 50 ft (15 m)

Fig 94 Geology of Woolacombe

Along this stretch of the N Devon coast, we see an alternation of sandy bays and rocky promontories. Similar alternations are common features of the British coastline and may be studied, for example, on the W side of the Lizard Peninsula and N and S of Swanage in Dorset. In such cases it will normally be found that the headlands can be correlated with the outcrop along the coast of rocks which are relatively resistant to erosion because of their chemical and/or physical properties. Such a correlation, considered in isolation, may lead to the conclusion that the relatively less resistant rocks have been eroded by the sea to form the bays—a process termed **differential erosion**.

Such a conclusion, however, is based only on the fact contained in statement (a) of question (2) above, and the other two statements suggest a different interpretation. Since we have seen (*p. 48*) that a rocky wave-cut platform and steep cliffs are evidence of active erosion, whilst sand and shingle beaches are the result of deposition, it is clear that under present conditions the action of the sea is to erode the headlands and deposit sand in the bays. In the case of Woolacombe, confirmation of this may be seen in the nature of the land behind the beach. This rises quite sharply from High Water Mark but is composed of sands, unconsolidated in places. In spite of its unresistant character, this slope shows no sign of active marine erosion. Promontory erosion and bay head deposition must inevitably lead to an eventual straightening of the coast line.

The diagrams (*Fig 95*) explain why as a result of **wave refraction** headlands are subject to active erosion, while waves breaking at the head of the bay have relatively less energy and are depositing material. It is also true that wave adjustment to the shape of the coast is seldom complete, so that in practice, a wave breaks on the headland a little before it breaks in the bay. Longshore drift will therefore tend to carry loose material from the headland towards the bayhead beach.

In the preceding paragraphs one sees the situation, frequently occurring in geomorphological studies, where two different lines of approach lead to contradictory conclusions. On the one hand, the coincidence of bays with less resistant strata suggests that the sea has detected this weakness and eroded the bays; conversely, study of present features indicates that deposition is now the dominant process in the bays. Can you suggest a way in which these ideas may be reconciled?

One approach to this problem is to consider whether the sea could first have eroded the bay to its present extent and yet now have altered its action to that of deposition. Alternatively it may be relevant to consider the possible effects of a change in sea-level. Is the origin of the bay the result of flooding of a lowland developed on the less resistant rock, with the headlands indicating the line of hills less affected by fluvial erosion? Different answers may be applicable to different examples of this type of coastline. There is always a danger in determining an origin for one physical landscape and then applying it too uncritically to another region which it superficially resembles. Every landform is unique and can only be fully understood after careful study, taking into account the general principles of the processes which have formed it, its geological structure, and the stage reached in its evolution.

FURTHER WORK

Practical

(1) Compare the features described above with the coastline N and S of Swanage (O.S. 1 : 63 360, Sheet 179, Bournemouth).

(2) Study maps and/or photographs of other areas of alternating bays and headlands. Look for evidence to support or refute the statement that 'headlands receive the major impact of wave attack and are subject to more active erosion than the intervening bays.'

Reading

Carter: *Land Forms and Life*, Section 33, pp. 248–254

Gresswell: *Beaches and Coastlines*, Ch. 4, pp. 61–63; Figs, pp. 68, 69 and 70

Steers: *Coastline of England and Wales*, Ch. 7, pp. 214–217

——: *Coastline of England and Wales in Pictures*, Nos. 65 and 66

Trueman: *Geology and Scenery*, Ch. 20, p. 274; Ch. 8, pp. 114–117

*Wooldridge & Morgan: *An Outline of Geomorphology*, Ch. XVI, pp. 211–236

Fig 95 Wave refraction

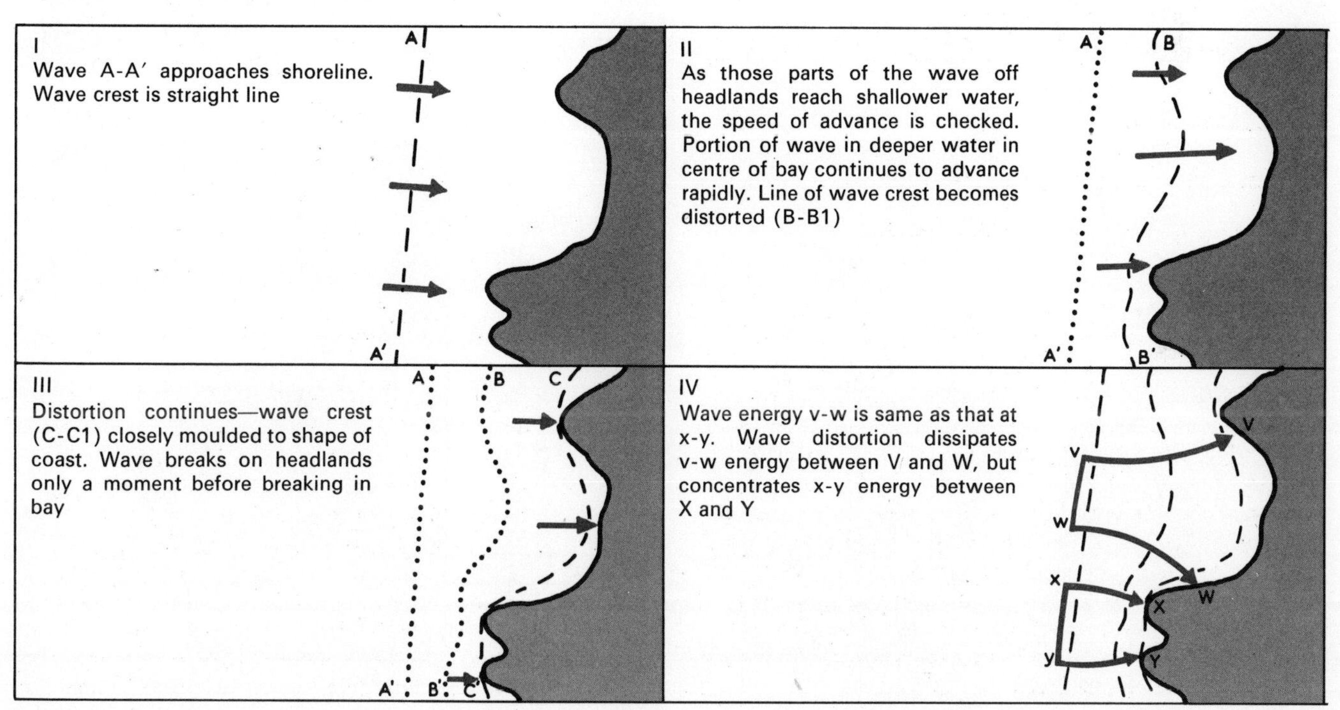

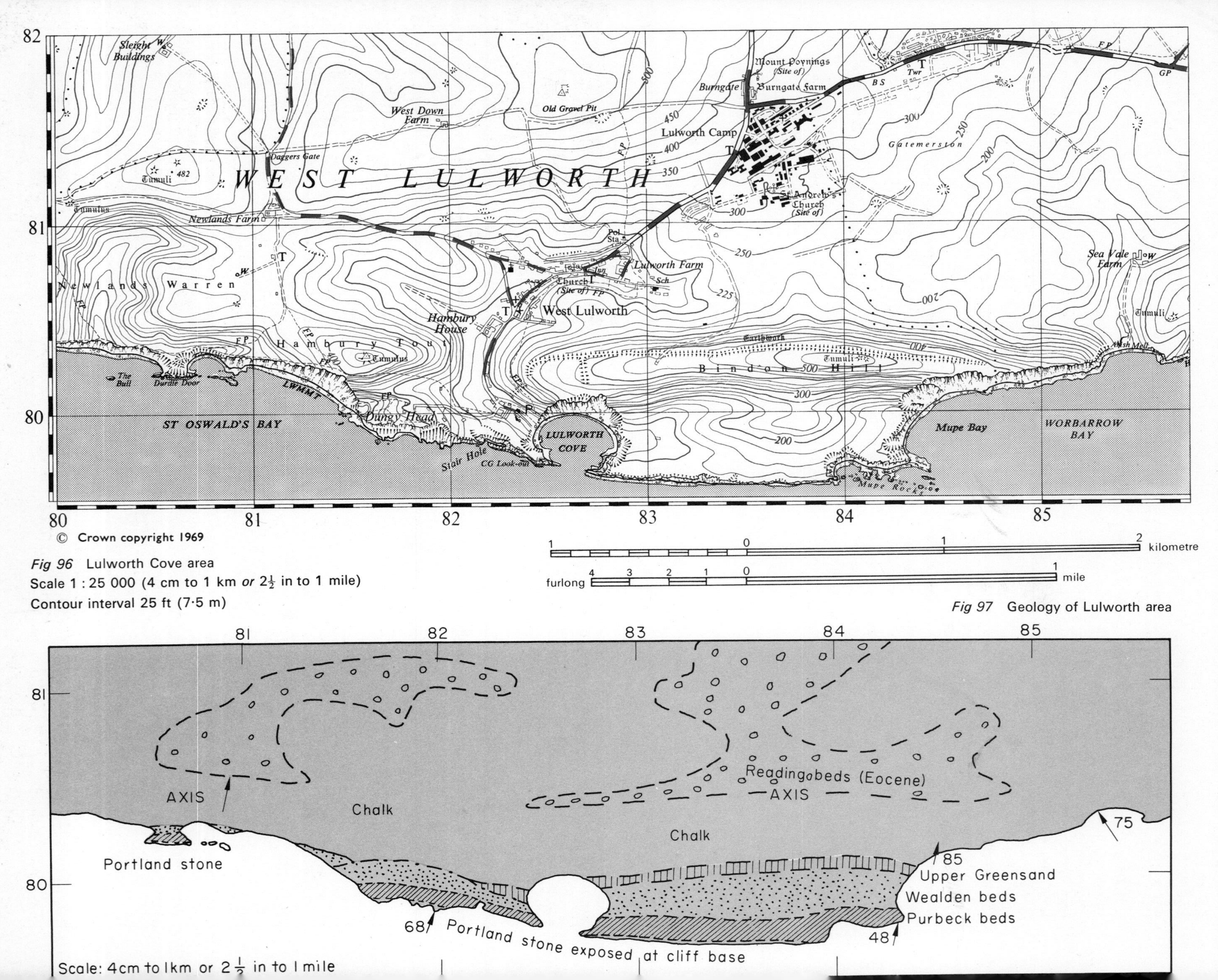

© Crown copyright 1969

Fig 96 Lulworth Cove area
Scale 1 : 25 000 (4 cm to 1 km *or* 2½ in to 1 mile)
Contour interval 25 ft (7·5 m)

Fig 97 Geology of Lulworth area

MARINE EROSION IN A REGION WHERE ROCK STRATA OUTCROP PARALLEL TO THE COAST

O.S. 1:63 360 (1.6 cm to 1 km *or* 1 in to 1 mile)
Sheet 178, Dorchester
O.S. 1:25 000 (4 cm to 1 km *or* 2½ in to 1 mile)
Sheet SY/88

In the Woolacombe area, studied in the preceding section, the form of the coastline closely reflected the rock structure. This is also true of the section of the South Dorset coast near Lulworth, but the resultant coastline in Dorset differs from that of North Devon for two main reasons. Firstly the rocks involved are different in character, and secondly the outcrops of the rocks are here arranged parallel to the coast, not at right angles to it.

MAP AND PHOTOGRAPH ANALYSIS

(1) Make a tracing of the geology of the area from *Fig 97* and superimpose this upon the relief map (*Fig 96*). What relationships between bed-rock and relief are suggested? (e.g. Which rock coincides with the highest ground?).

(2) Consider this relationship more precisely by constructing a section along Easting 83. Indicate on your section the area in which various rocks outcrop.

(3) Locate the line of your section on the aerial view of Lulworth Cove (*Fig 98*). Notice how aerial views tend to 'flatten' a landscape.

(4) Make a tracing from *Fig 98* and on it show the location of cliffs, beaches, steep slopes and the pattern made by waves as they enter the cove. Add an arrow to show North. Mark also the eastern end of Stair Hole which is just visible at the bottom of the photograph. Refer to the geological map and indicate the various rock outcrops on your tracing. Explain why small vessels may usually be moored safely within the cove.

(5) What evidence is there from (a) the photograph, (b) the map that the rocks in this area are not horizontal?

The geological map (*Fig 97*) shows by the narrowness of the outcrops along the coast, and by the arrow numbered 68 (giving the angle of dip of the rock in degrees

Fig 98 Aerial view of Lulworth

Fig 99 Durdle Door promontory from the east. The arch of Durdle Door is on the far side of the promontory

from the horizontal), that these rocks are very steeply inclined. This is clearly shown on the photograph (*Fig 100*) which also suggests the way in which intense pressures resulting from past earth movements have crumpled and contorted the strata which were once horizontal sediments beneath the sea. The rock actually forming the coastline at this point is relatively resistant Portland Stone; the crumpled strata forming most of the cliff in the centre of the view are the Purbeck Series which are also resistant to erosion to a considerable degree, but which have here been weakened by intense folding. Further inland, on the left of the photograph are the much less compact clays and sands of the Wealden Beds.

At Stair Hole, which lies 275 m W of the entrance to Lulworth Cove, one may see the first stage of coastal erosion in this area. The waves have opened up joints in the Portland Stone forming a narrow cleft and also two arches through which waves can be seen breaking in *Fig 100*. Through these points of weakness the sea has encroached and been able to erode some of the crumpled Purbeck Beds until it now washes at high tide against the Wealden clays and sands. There is clear indication of soil creep and mass slumping of these clays towards the sea, and this unconsolidated material is easily carried away by the action of the waves.

Close study of Stair Hole, particularly in the field, leaves little doubt as to the mode of origin of Lulworth Cove itself. A weak point in the coastal 'wall' of Portland Stone provided an opening to the erosive action of the waves or alternatively the opening may have been formed by river erosion. With the removal of the Purbeck Beds, and the slow enlargement of the original opening, the sea was able to exploit the relatively unresistant Wealden Beds, eroding these more quickly than the rocks of the opening. The back wall of the cove is now cut into the Chalk outcrop, though the height of the chalk ridge and nature of this rock must necessarily be a check to further enlargement of the cove.

It is reasonable at this point to question whether waves can occur within the cove to achieve the erosion suggested. *Fig 83* and *98* shows how waves are deflected upon entering the cove. The wave energy passing through the cove entrance is distributed all round the shores of the cove, but even so under stormy conditions the waves within the cove are quite sufficient to have an erosive effect, especially upon rocks of limited resistance, and also to remove debris carried down to beach level by slumping in the Wealden Beds.

Returning to Stair Hole, one can imagine how its gradual enlargement could lead to another 'Lulworth' Cove and that these two coves could become joined to leave the site of the coastguard look-out as a rocky island, which would itself be slowly eroded away. This is the process by which Worbarrow Bay, east of Lulworth, was formed. The broader outcrop of Wealden Beds has permitted deeper penetration by the sea here. W of Lulworth the coastline is being driven back in the same manner. Lines of rocks jutting just above the sea (*Fig 99*) and parallel to the coast either side of Durdle Door show the line where Portland Stone outcrops. Durdle Door itself (*Fig 101*) is a natural arch, parallel to the coast, carved in nearly vertical Portland Stone.

Coastal erosion in this area thus passes through a number of recognizable stages:

(1) A relatively straight coast with cliffs of Portland Stone.

(2) The opening up of weak points in the Portland Stone and Purbeck Beds to form inlets and coves.

(3) The enlargement of these coves by the erosion of the less resistant Wealden Beds.

(4) The gradual amalgamation of enlarging coves until the coastline is formed by the Chalk outcrop with isolated promontories and stacks of Portland and Purbeck Beds—the remnants of the land between the coves.

(5) The eventual removal of these irregularities leaving a straight coastline backed by chalk cliffs.

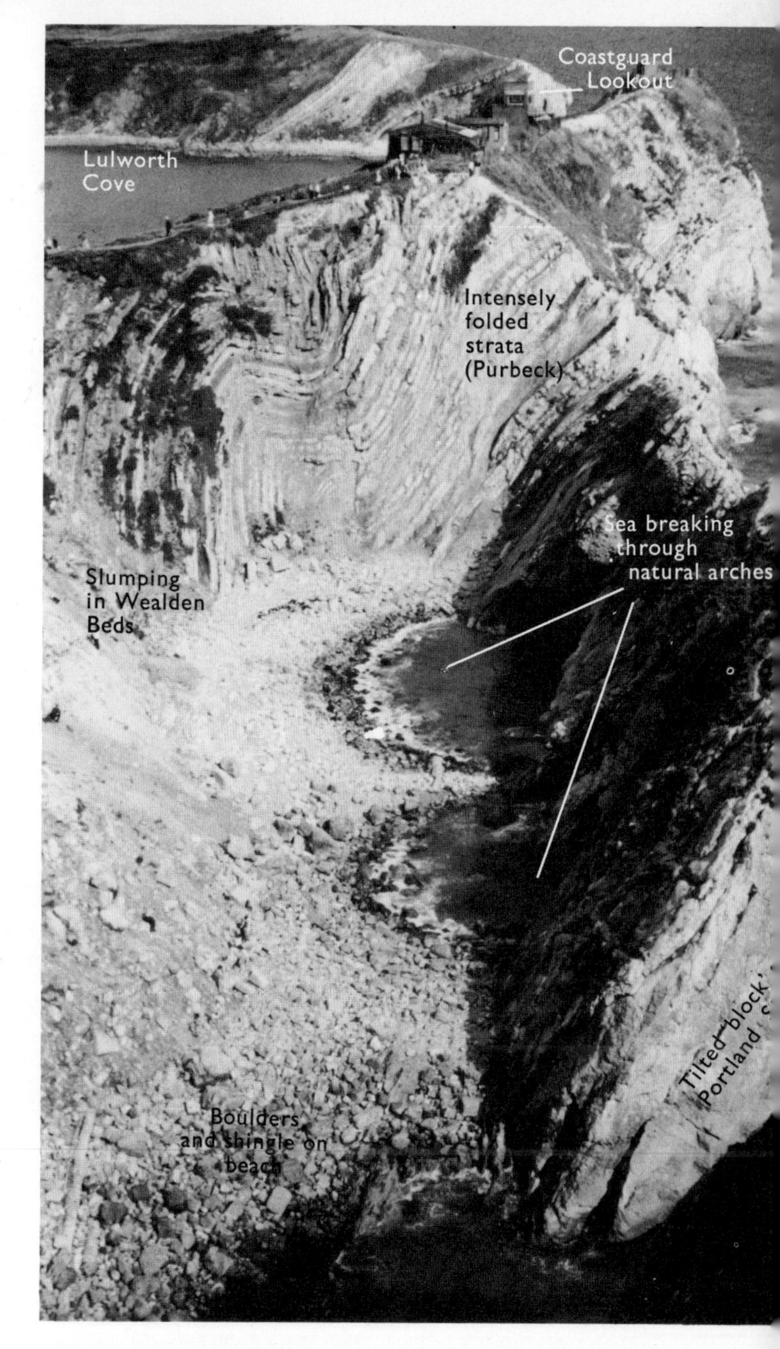

Fig 100 Stair Hole from the west

FURTHER WORK

Practical

(1) Use the photograph (*Fig 99*) as the basis for a field sketch, indicating the significant geological outcrops. Annotate your sketch with sufficient marginal notes, etc. to summarize the stages in coastal erosion described above.

(2) Explain carefully the probable relationship between the small caves of Stair Hole (*Fig 100*) and the natural arch of Durdle Door (*Fig 101*, overleaf).

(3) Using tracings from *Fig 96* and *97* as a basis, draw a series of sketch maps to show the stages in the evolution of this stretch of coast: (a) the coastline of unbroken Portland Stone (note a small river flowing into Lulworth Cove from the N.W. which may once have been larger)

(b) early breaches in the Portland Stone forming coves in Worbarrow Bay and either side of Durdle Door promontory,

(c) the present stage,

(d) a future situation with Stair Hole enlarged to a cove and Durdle Door, etc. removed,

(e) a still further stage with the coastline coinciding with Chalk for its whole length.

Reading

Gresswell: *Beaches and Coastlines*, Ch. 4, pp. 63–66

Hardy & Monkhouse: *Physical Landscape in Pictures*, p. 13

Horrocks: *Physical Geography and Climatology*, Ch. 9, p. 151

Steers: *Coastline of England and Wales*, Ch. 8, pp. 281–284

——: *Coastline of England and Wales in Pictures*, Nos. 26, 27, 28, 29

Trueman: *Geology and Scenery* Ch. 8, pp. 117–120

Fig 101 Durdle Door *(see page 67)*

Study B8

Fowey Estuary, Cornwall

DROWNED COASTLINES

O.S. 1:63 360 (1.6 cm to 1 km *or* 1 in to 1 mile) Sheet 186, Bodmin and Launceston

The map extract shows the lower valley and estuary of the Fowey River on the S coast of Cornwall. At its mouth the estuary is about 300 m wide and extends inland for almost 8 km with a width of between 300 and 200 m. The river is tidal to 1 km N of the town of Lostwithiel. Within the estuary are some sharp bends, notably about 128528, and the main channel is joined on its left (E) bank by three similar though smaller estuaries, each of which has a small stream flowing into it. In the smaller estuaries and at the head of the Fowey Estuary itself there is evidence of alluvial deposition and the main river channel can be seen to meander from side to side of the estuary between mud flats exposed at low tide.

Except where the frequent small tributary valleys join the estuary, the land rises fairly steeply from sea-level to about 100 m (300 ft contour). In some places this rise is achieved in less than 0.3 km (e.g. 123535), in other places (e.g. 118575) the 300 ft contour is almost 0.6 km from the water's edge. Above 300 ft (100 m) the contour spacing tends to increase showing that many hills are fairly flat-topped at about this height. The whole area is very dissected by steep-sided stream valleys, with one or two hill tops rising a little over 125 m (400 ft contour).

The S coast is rocky with cliffs rising to 100 m (300 ft contour) in places, minor bays and inlets, an almost continuous narrow wave-cut platform and no evidence of sandy beaches.

From the preceding description it is possible to select the following points to make a concise analysis of the land forms of this area. 'An undulating plateau surface between 100 m and 125 m, is dissected by a well-developed system of steep sided stream valleys, the lower sections of the main valleys having been flooded by the sea to form long narrow inlets.' This is a description of a **ria** type of coastline which is a common feature of much of Devon and Cornwall. In *Study A11* and elsewhere, reference has been made to evidence for a relative lowering of sea-level, but, in this instance, there is clear evidence for the reverse process. The **drowning or submergence of coastlines** is indeed a very common feature in the British Isles and elsewhere consequent upon a rise in sea-level of some 100 m which has taken place during the past twenty thousand years (*Study B5*). On all but the very flattest coastlines, any rise of sea-level will result in the flooding of valley floors and will thus produce an indented coastline. Where the lower portions of valleys are narrow and steep sided the resultant inlets will be long and narrow; if a low lying coast with broad river valleys is flooded the inlets will be wider with gently sloping shores such as the estuaries of the Stour and Orwell in Suffolk and the Blackwater in Essex.

The form of a submerged coastline may also reflect the geological structure of the area. The inlets of SW Ireland, for example, have been formed by the drowning of vales eroded in carboniferous strata preserved in synclines while the intervening ridges are of resistant Devonian rocks. The SW–NE alignment of the inlets thus emphasizes the trend of anticlinal and synclinal axes of folding (*Fig 103*).

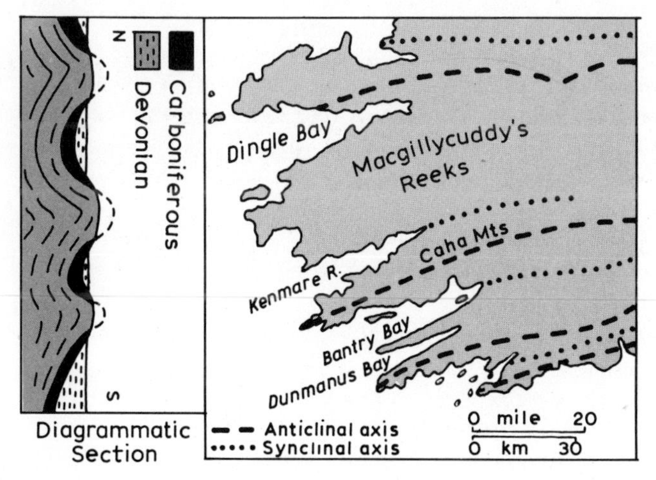

Fig 102 South West Ireland

Where structural trends are parallel to the coast, submergence may produce a very different coastal form such as that of the NE coast of the Adriatic Sea (*Fig 104*). The mainland shows the dominant NW–SE structure in parallel longitudinal valleys separated by ridges rising to 225 m (spot height 778 ft). Off shore, further such valleys have been inundated by a rising sea-level leaving the ridges as long narrow islands parallel to the coast. It is from this area that a coastline of this type gains its name of a **Dalmatian Coast.**

As soon as an indented coastline has been formed it is subject to modification by alluvial deposition and by the processes of marine erosion and deposition. Headlands come under the most rigorous attack of waves and are thus characterized by cliffs, stacks and a rocky wave-cut

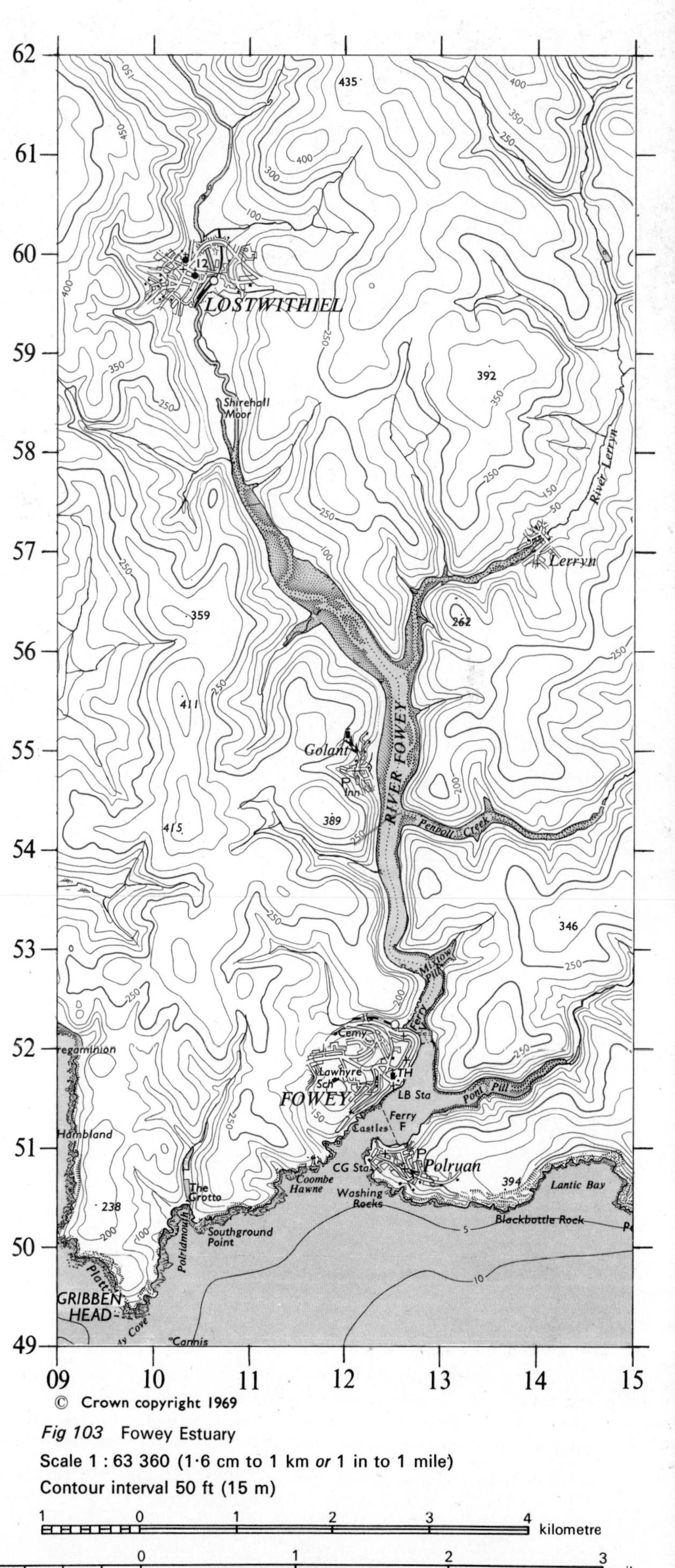

Fig 103 Fowey Estuary

Scale 1 : 63 360 (1·6 cm to 1 km *or* 1 in to 1 mile)

Contour interval 50 ft (15 m)

platform. At the same time alluvium brought down by rivers and the debris of marine erosion transported by longshore drift collects to form bay head beaches. In the case of long narrow inlets, such as the Fowey River, the infilling at the head of the estuary may be due almost solely to alluvium supplied by the river. Some deep inlets become partly or completely blocked at their mouths by bay mouth spits formed under the influence of longshore drift—for example, the Exe and Teign estuaries in Devon, the Dovey estuary in West Wales, or Poole Harbour in Dorset (*note* also the Beaulieu River in Hampshire—*Study B4*). Such partial obstruction of the estuary restricts the action of tidal currents and waves in redistributing river borne alluvium and thus encourages the infilling of the estuary. A drowned coastline, originally deeply indented, is thus gradually straightened by the combined effect of sedimentation filling the inlets and wave attack reducing the promontories.

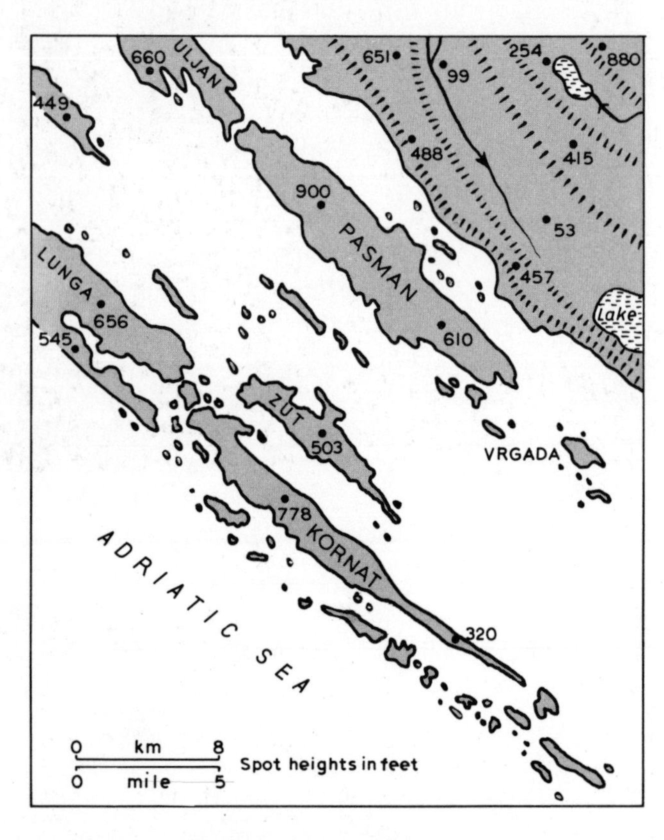

Fig 104 Dalmatian Coast, Yugoslavia

FURTHER WORK

Practical

(1) Use the 1:63 360 map of any fairly straight coast, and sketch the coastline which would result from a rise in sea-level of (a) 50 ft, (b) 100 ft.

(2) Note task 2 following *Study C6* and task 1 following *Study E5*, both of which require study of the Fowey Estuary.

(3) The first three paragraphs of this Study comprise a descriptive account of the area shown on the O.S. map extract. From it alone a map could be constructed which would not differ in many essentials from the original. A very valuable exercise is for two students to write such descriptions from different maps, to exchange descriptions and attempt to draw a sketch map based on the description, and finally to compare their maps with the originals.

(4) Draw sections across the Fowey estuary to show its characteristic features and draw an annotated sketch map of the area. What are the effects of such a coastline upon the human geography? (Refer to O.S. 1:63 360, Sheet 186, for the features of human geography.)

(5) From O.S. 1:63 360, Sheet 150 (Ipswich), write a descriptive account of the estuaries of the Orwell and Stour. What diagrammatic data (sections, etc.) would you use to supplement your description?

(6) Consider how the following factors will affect the rate at which a drowned inlet will become filled by sediment: Velocity and volume of rivers entering the inlet; nature of rocks in the basins of these rivers; depth of submergence; steepness of cross-section of inlet; tidal range.

Reading

Carter: *Land Forms and Life*, Section 30, pp 232–237; Section 31, pp. 238–241; Section 32, pp. 242–247
Gresswell: *Beaches and Coastlines*, Ch. 4, pp. 52–60
——: *Physical Geography*, Ch. 28, pp. 383–384
Hardy & Monkhouse: *Physical Landscape in Pictures*, pp. 78–80
Horrocks: *Physical Geography and Climatology*, Ch. 9, pp. 146–147
Monkhouse: *Principles of Physical Geography*, Ch. 10, p. 254
——: *Landscape from the Air*, p. 41
Scovel: *Atlas of Landforms*, pp. 132–133
Sparks: *Geomorphology*, Ch. 9, pp. 211–218
Steers: *Coastline of England and Wales in Pictures*, No. 44
Strahler: *Physical Geography*, Ch. 27, pp. 417–422; Ex. 1, p. 429
Trueman: *Geology and Scenery* Ch. 21, pp. 283–4

Fig 105 The Rhône Glacier

The Rhône Glacier, Switzerland

Study C1

A LANDSCAPE DURING GLACIATION

Landeskarte der Schweiz 1:50 000 (2 cm to 1 km *or* 1¼ in to 1 mile)
Sheets 510 and 511

Every student in Britain can easily observe the flow of rivers and streams, and most have at least occasional opportunity to study waves breaking on our coastline. The majority of readers will not have seen a glacier or ice sheet, and their knowledge of permanent snow-fields, ice forms and action must necessarily be derived from second-hand sources. Yet the land forms of Britain, as of many other temperate lands, have been enormously influenced by the erosional and depositional processes which occurred during the Pleistocene Ice Age, and such features have in some areas suffered only slight modification by fluvial erosion processes in the last eleven thousand years. And this influence of **glaciation** is not limited to the mountain areas but extends to much of the English lowlands with the deposition of glacial debris, modification of drainage patterns, remnants of glacial lake deposits and so forth. Even south of the maximum limits of the ice sheet, approximately from Bristol to the Thames, the effects of a Tundra-like climate during the Ice Ages must be taken into account. Features, occurring on the fringes of the ice sheets, are known as **periglacial.**

The above photograph and *Fig 106* illustrate one part of the Swiss Alps where ice is at present an active agent of landscape formation. At these high altitudes the majority of precipitation falls as snow, not all of which melts during the cool summers. As a result the first snows of winter fall on to old snow which has survived the summer and in this way the amount of snow increases annually. It can be seen from the mountains in the background of the photograph that some slopes are too steep for snow to cling to, and from these areas and from steep snow slopes, masses of snow fall as avalanches

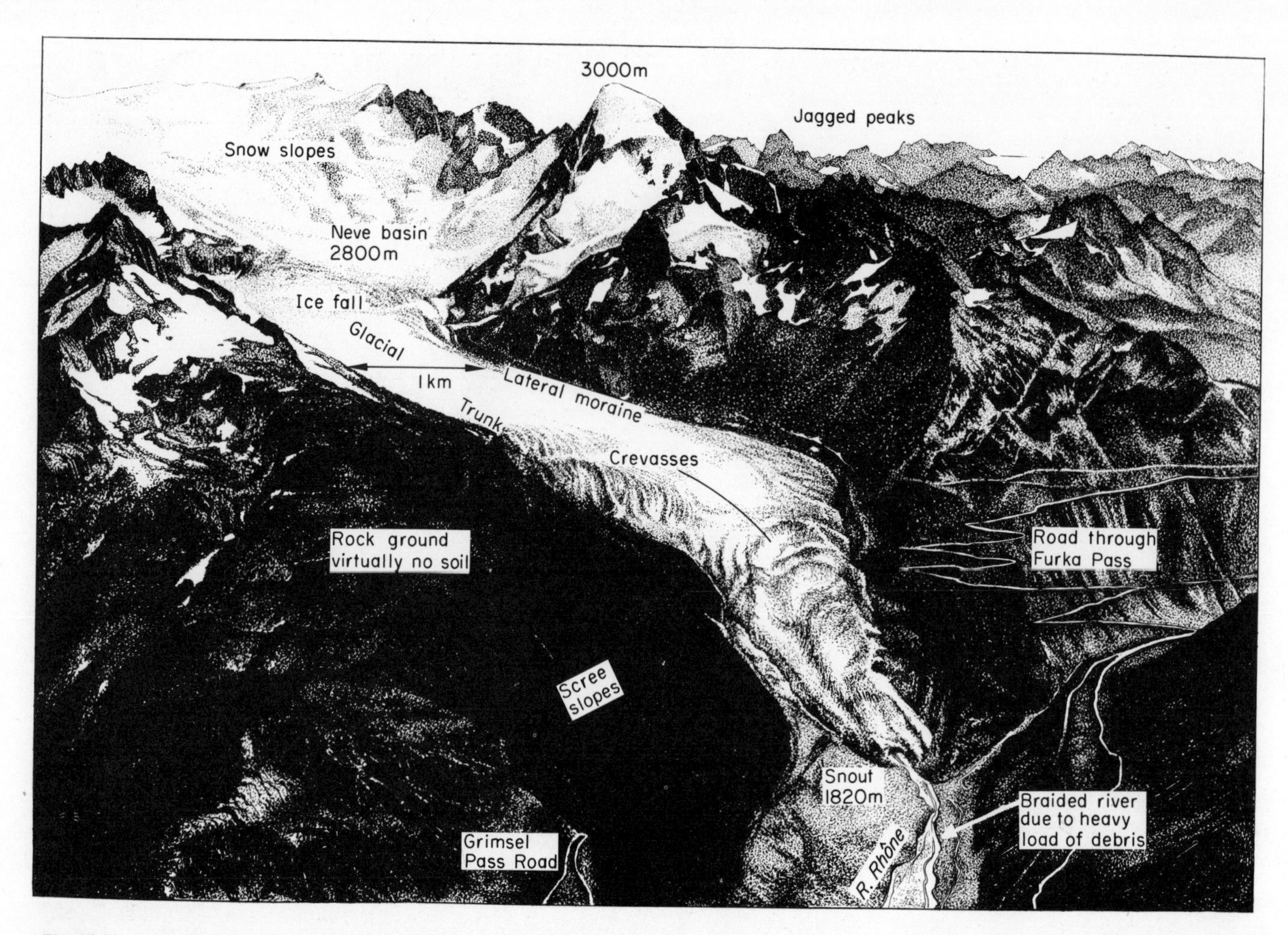

Fig 106 Annotated sketch of the Rhône Glacier (drawn from *Fig 105*)

to accumulate in the hollows below. As the depth of snow in such hollows increases, ever greater pressure is exerted on the deeper layers of snow until it is compacted into ice in which some air is trapped. This white ice, called **névé** or **firn**, may be compared to the formation of icy footprints in light, powdery snow—also the result of pressure. Further compacting of the névé excludes the trapped air and leaves the 'blue' glacier ice.

The processes by which the 'solid' ice so formed is induced to move are complex, probably involving both a form of **plastic flow** between crystals and the process of **regelation** by which ice crystals may temporarily melt under pressure, flow a minute distance as a liquid and then refreeze because some of the excess pressure has been relieved. Not only are the processes of movement complex, but the motion of ice when it has collected in a basin involves movement in a number of directions. Unevenness of the surface of a névé basin leads to variations in pressure, which in turn impart a horizontal rotational movement to the ice, and at the same time ice is squeezed upwards over the lip of the hollow and begins to move along a pre-existing river valley, slowly filling and deepening it. In summer a wide and deep crevasse—the **bergschrund**—may open between the upper edge of the ice in a névé basin and the steep mountain slope behind. Consequent upon the rotational grinding movements of ice in the névé basin, and the rock material contained within it, aided to some extent by frost action where the mountain side is exposed to atmospheric changes by the opening of the bergschrund, the basin is gradually deepened and enlarged to form a **cirque** (*see Study B2*). This enlargement may eventually leave only a steep narrow ridge with frost shattered upper slopes separating one cirque from its neighbours.

Immediately below the edge of the névé basin it is common to find, as in the Rhône Glacier (illustrated), a sharp increase in gradient. The various processes inducing glacial flow are unable to cope with this change in angle, so that the ice here is found to be fractured into large blocks, separated by deep, narrow clefts, or **crevasses**, which can be distinguished in the photograph. Although stresses normally produce a complicated pattern of crevasses at such places, and indeed generally along the glacier's course, the dominant crevasses can be seen to form arcs curving across the glacier with their convex side facing upstream. The explanation of this is

why not facing downstream

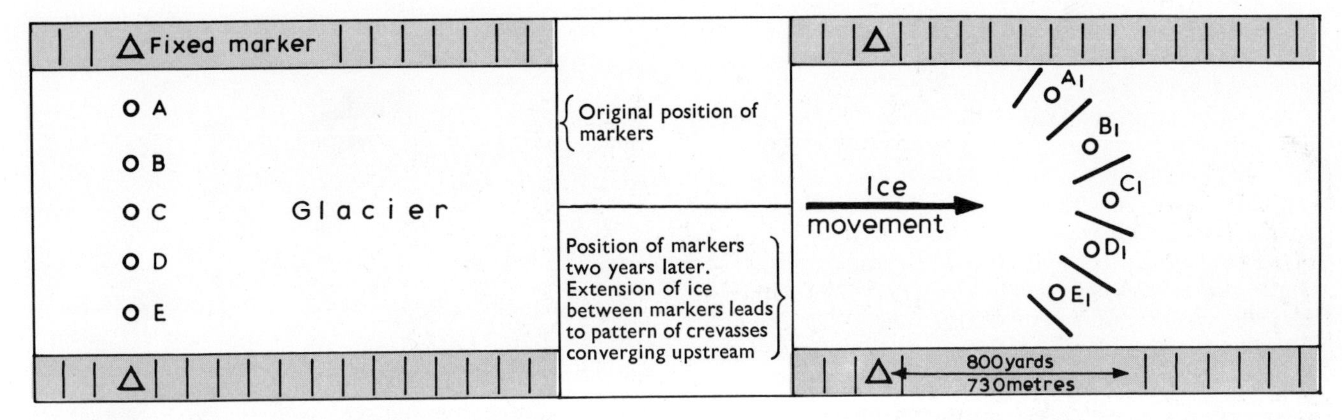

Fig 107 Movement at surface of glacier

that ice in contact with the valley sides is retarded by friction and therefore movement is more rapid in mid-stream. Study of *Fig 107* will show that this must result in extension of the ice between points A and B, B and C and so on, and this tension is relieved by the formation of cracks and crevasses. When the ice surface is once broken in this way it does not easily coalesce again so that most glacier surfaces are rugged and broken, and only appear smooth when blanketed by further snow fall.

In considering the movement of glaciers one must realize that the speed of motion involved is of a different order from that of rivers. The most rapid glacial movement recorded is about 18 m per day, while 1 m or so per day may be taken as a more normal figure. As with a river, the speed of movement is a function of gradient and volume, but one must also consider for glaciers such factors as the rate of supply of ice, and the temperature. Since for comparable gradients water can flow at 100 000 times the speed of ice, it follows that for any given drainage basin the cross-section of a glacier must be enormously greater than the cross-section of the river which might exist under warmer conditions. A glacier therefore will largely fill the valley in which it flows, sometimes to depths of a 1000 m or more, and widths of up to 5 km, and sometimes wider. The Rhône Glacier is approximately 1–1.5 km wide, and its length from the peaks on the left skyline to the foreground is 8 km.

a) By contrast with the curving nature of rivers in mountainous areas, it can be seen that the **trunk** (main course) of the glacier is relatively straight. This feature of glaciers is due to the relatively solid nature of ice which obviously cannot easily adapt itself to flow round the spurs of any pre-existing river valley; hence when we consider the landscape left after the ice has melted (*Study B3*) we must look for indications of a straightening of the valley.

b) Two other significant aspects of glacial flow follow from the essential difference that ice is a solid and is not governed by laws of flow which relate to liquids. In the first place ice can, if the pressure is sufficient, be pushed up hill. The glacial trunk can thus sculpt hollows along its long profile. c) The other result of the solid nature of ice concerns the relationship between main glaciers and their tributaries. In the case of river tributaries, any down-cutting of the main stream steepens the gradient and increases the erosive power of the tributary. The long profile of the tributary therefore quickly becomes adjusted to the point of confluence (*Fig 108a*). In the case of

Fig 108 Tributary junctions of (a) rivers and (b) glaciers

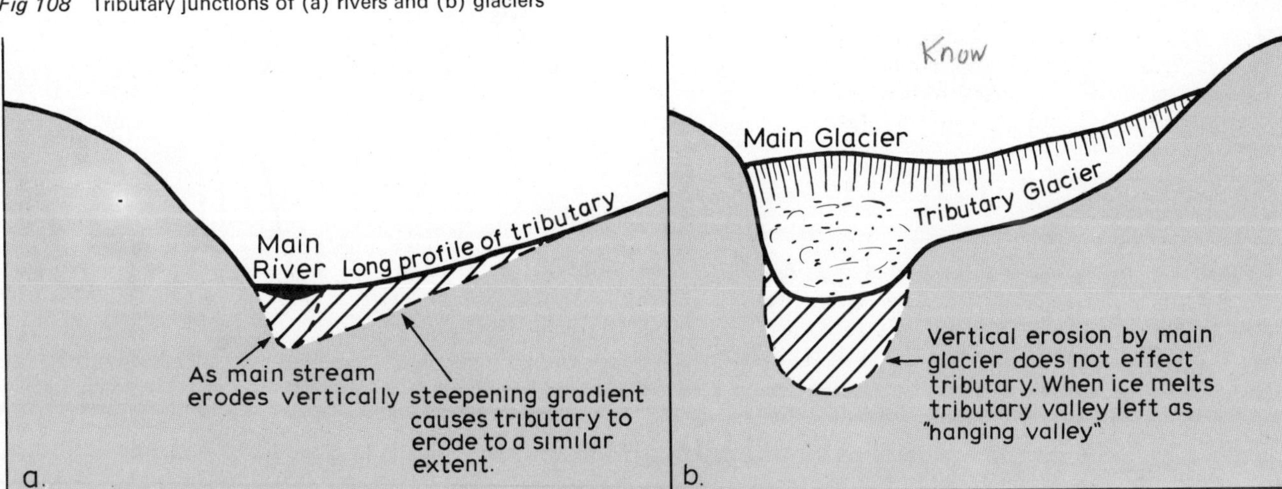

glaciers this simple relationship does not apply, so that a sharp difference may exist between the bed of a main glacier and its tributary (*Fig 108b*).

The photograph shows that the Rhône Glacier moves downward (from 2800 m in the névé hollow to 1820 m in the foreground) to below the level of permanent snow. At any point along its course its size is determined by the balance between the supply of ice from above and the loss by thawing. As progressively lower levels are reached the mean temperatures increase and so too does the amount of thawing, until at the end, or **snout**, of the glacier the last of the ice supply is thawed. It follows that throughout the final sections, there is a steady increase in the amount of melt water which flows through ice tunnels in, or at the base of, the glacier. In the case of the Rhône Glacier this melt water emerges from an ice cave at the snout which is the source of the River Rhône.

When the supply of ice increases, as for example following many years of increased snow fall, or when the amount of thawing decreases consequent on a lowering of mean temperature, the glacier will **advance**, i.e. ice will gradually extend further down the valley. Conversely, a decrease in snowfall or increase in mean temperature will lead to a **melting back,** or **retreat**, of the snout to a higher level. The latter is occurring at the present time, and the Rhône Glacier's snout is now at an appreciably higher level than it was fifty years ago. It should be noted that the term 'glacial retreat' does *not* imply that the ice itself does not continue to move downwards, but that the snout is found at progressively higher levels. The last $1\frac{1}{2}$ million years have witnessed a number of advances and retreats of ice formations in the Northern Hemisphere and a study of *Fig 110* will show that we may well be living in yet another **inter-glacial** epoch.

If we now consider the rocky slopes above the glacier, it will be clear that these are subject to intense weathering by frost action, and upper slopes are rendered jagged and sharp by this process. Rock fragments so formed accumulate in scree slopes; and some of this debris falls on to the sides of the glacier to be carried forward as a band of **lateral moraine.** If two glaciers join, two of the lateral moraines will unite and move down the centre of the combined glacier as a **medial moraine** (*Fig 109*). Some of this lateral and medial moraine will be carried by melt water or ice movements into the body of the glacier where it joins with rock fragments eroded by the ice from the valley sides and floor to form **englacial moraine.** Finally all this morainic material reaches the snout to be deposited as **terminal moraine** or redistributed by melt water streams. So-called **Mountain** or **Valley glaciers** such as the Rhône Glacier occur today in many of the world's great mountain ranges from high latitudes to the Equator. Since their erosive power is confined to the valleys through which they flow, their most characteristic effect is to increase the **amplitude of relief**—that is the difference in altitude between the peaks and the valley floor. In still colder regions, however, the volume of ice increases until it more or less completely buries the land. Such a formation, extending perhaps over millions of square kilometres, is an **ice sheet.** Today there exist

Fig 109 Inner lateral moraines (L1 & L2) join to form medial morain (M)

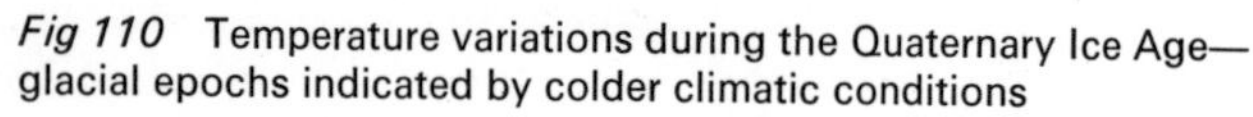
Fig 110 Temperature variations during the Quaternary Ice Age—glacial epochs indicated by colder climatic conditions

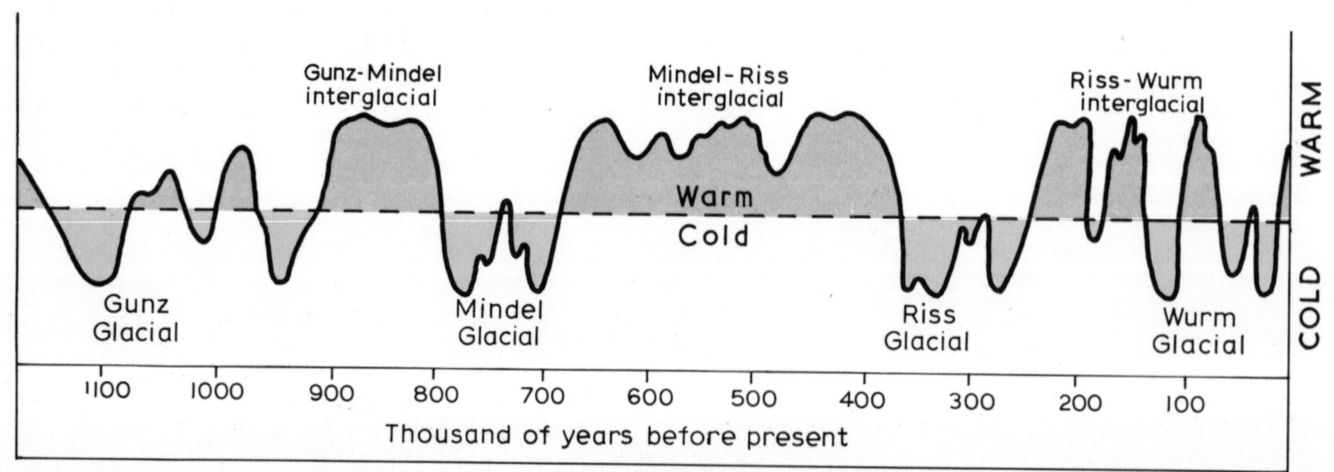

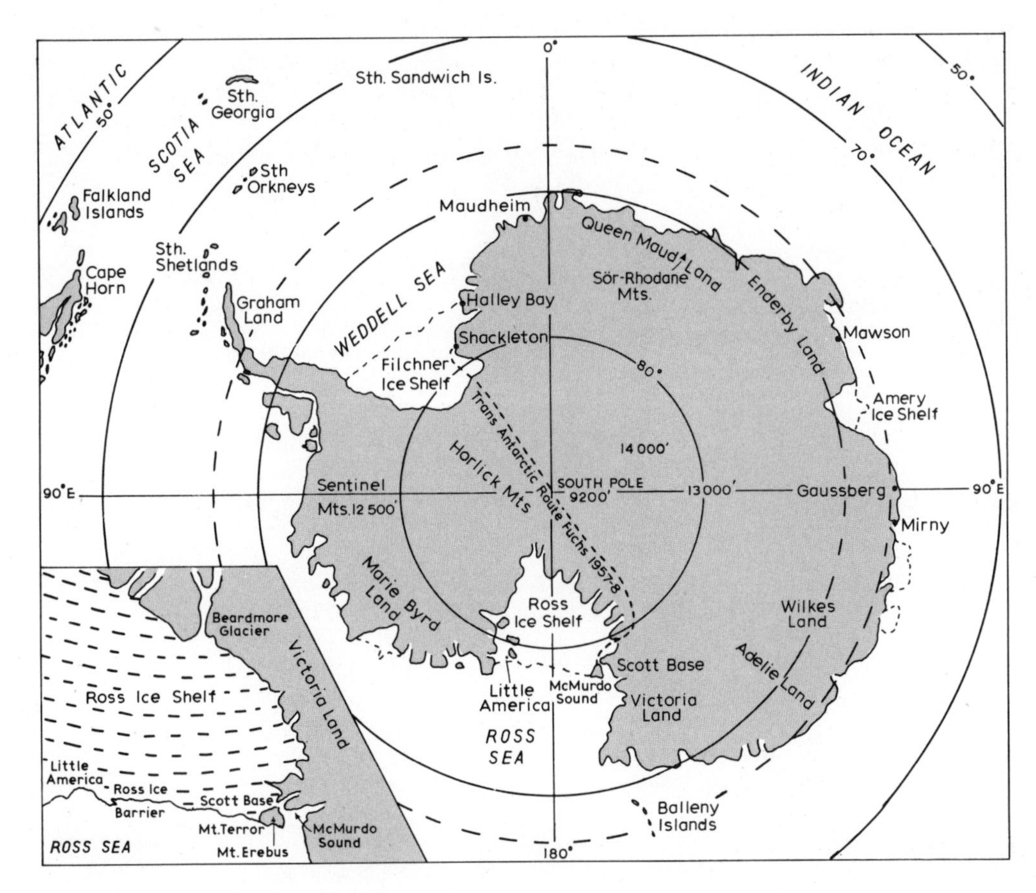

Fig 111 Antarctica

two such ice sheets, Antarctica and Greenland, with a number of more limited ice-covered uplands known as **ice caps**—e.g. in Iceland, Spitzbergen, and the islands of the Canadian Arctic. During the Pleistocene period ice sheets extended southward many times to cover Canada and the northern United States, much of the British Isles and Northern Continental Europe.

The Greenland and Antarctic Ice Sheets of the present day cover approximately 1 300 000 km^2 and 9 000 000 km^2 respectively, and in places exceed 3000 m in thickness. The surface of the ice at such points may be less than 3000 m above sea-level and it therefore follows that the base of the ice rests on land which is below sea level. It is the enormous weight of ice which is partially responsible for the low altitude of land in the centre of Antarctica and Greenland; during the Ice Ages the land surface of Britain, Western Europe and North America would have been similarly depressed and the effect of its slow rise following the removal of ice has been referred to on p. 33 and is further dealt with in *Study C6*.

Whereas the erosive and depositional forms of valley glaciers are largely confined to the valleys through which they move, an ice sheet mantles and modifies almost all the land surface. Only occasionally, and usually on the margins, do high peaks project through the ice sheet as **nunataks** (*Fig 113*). With these exceptions the land is rounded and smoothed by the ice; soil and rock debris is transported towards the ice margins; rock surfaces are polished and scratched (or striated). In fact the erosive processes beneath an ice sheet are the same as those beneath a valley glacier, but whereas the latter deepens only a narrow trough through a mountain zone an ice sheet covering the whole land mass has the overall effect of reducing the pre-glacial amplitude of relief.

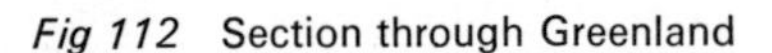
Fig 112 Section through Greenland

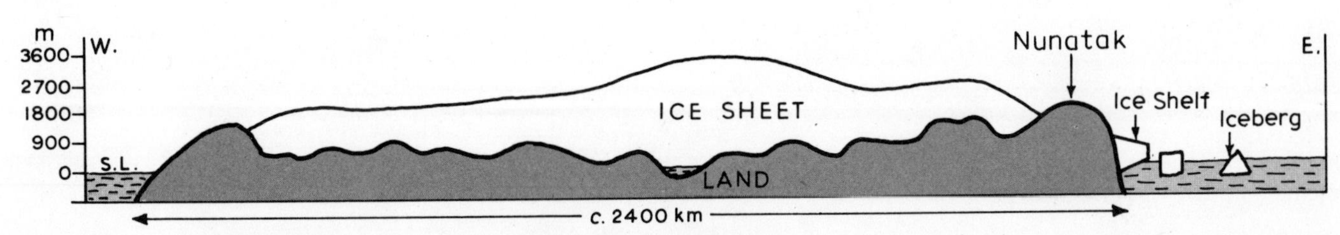

FURTHER STUDY

Practical

(1) Annotated sketches should be drawn based on photographs of glacial scenery such as those which appear in Monkhouse *Landscape from the Air* or Holmes *Principles of Physical Geology.*

(2) Study any Alpine sheet of the 1:50 000 Landeskarte der Schweiz. First notice how the very beautiful representations of mountain scenery are achieved—i.e. cartographic appreciation. Then draw a sketch map of a glacier and annotate as many as possible of the features mentioned in the above chapter.

(3) If possible, view the film *Evidence for an Ice Age* (Rank Film Library) and summarize the evidence which is presented for past glacial epochs. Excellent glacial scenes occur in the films *Ascent of Everest* (Rank) and *Crossing of Antarctica* (Petroleum Film Bureau).

Reading

Anderson: *Splendour of Earth,* pp. 315–327

Dury: *The Face of the Earth,* Ch. 12, pp. 137–147; Ch. 13, pp. 148–163

Gresswell: *Glaciers and Glaciation,* Ch. 1, pp. 9–13; Ch. 8, pp. 111–128

——: *Physical Geography,* Ch. 19, pp. 219–240; *Ch. 22, pp. 275–296

Hardy & Monkhouse: *Physical Landscape in Pictures,* pp. 52–55

Holmes: *Principles of Physical Geology,* Ch. XX, pp. 619–641; Figs, pp. 698, 699, 702–703, 706–707

Horrocks: *Physical Geography and Climatology,* Ch. 6, pp. 90–98

Monkhouse: *Principles of Physical Geography,* Ch. 8, pp. 173–188

——: *Landscape from the Air,* pp. 24, 25

Scovel: *Atlas of Landforms,* pp. 104–105

*Sparks: *Geomorphology,* Ch. 12, pp. 264–274

Wooldridge & Morgan: *An Outline of Geomorphology,* Ch. XXII, pp. 330–334

Fig 113 Part of the Antarctic Ice Sheet, Palmer Coast, Grahamland—showing nunataks

CORRIES, ARÊTES AND TARN

O.S. 1:63 360 (1·6 cm to 1 km *or* 1 in to 1 mile)
Lake District (Tourist Edition)
O.S. 1:25 000 (4 cm to 1 km *or* 2½ in to 1 mile)
Sheet NY 31

The English Lake District is a region much favoured by ramblers and rock climbers of a wide range of skills and ability. One very popular ascent which can be undertaken by those with little experience, though always requiring common sense precautions relative to clothing, footwear and possible weather developments, is to set out from Patterdale on Ullswater and climb Helvellyn by way of Striding Edge. For the first 3 km W of Patterdale the path climbs steeply but smoothly over grass and bracken covered hillside from an altitude of 150 m (500 ft contour) at Patterdale to almost 700 m (spot height 2283 ft) by the time the summit of Grisedale Brow is reached at G.R. 360155 (*Fig 115*). From 450 m (1500 ft contour) upwards, the ground becomes progressively more stone littered, vegetation is more sparse and **scars**, or minor cliffs of bare rock show on the steep hillside. At G.R. 360155 one is standing at the top of a long steep-sided ridge or spur which juts eastward from Helvellyn towards Ullswater and which separates the valleys of Red Tarn Beck and Grisedale Beck. The latter stream is 450 m (1500 ft) below the crest of Grisedale Brow and only some 1100 m (1200 yd) away in horizontal distance. (What is the average gradient of this slope? What is the angle of slope? Use the map extract to describe the view down into Grisedale.)

From this point onwards, to the summit of Helvellyn the path lies over bare rock and the ridge along which one climbs becomes progressively narrower. The steepening is first noticed on the S, or Grisedale, side where the ground falls away almost vertically from the ridge top and below this cliff-like edge is a straight slope of stones and boulders—scree (*Fig 114*). To the N also steep scree slopes fall to the valley of Red Tarn Beck, and looking across the lake one sees the comparable rise to Swirral Edge. The area of land immediately round Red Tarn appears relatively flat, and it can be seen that on leaving the lake the water of Red Tarn Beck tumbles steeply into a valley at a somewhat lower level.

By the time High Spying How (G.R. 350149) is reached, the crest of the ridge has narrowed to less than 3 m in places with cliff-like drops on both sides and massive scree slopes below, and in places the rambler must become something of a 'scrambler' in order to negotiate the irregular track. This ridge is the **arête** known as Striding Edge. Striding Edge and Swirral Edge are two 'arms' of high ground stretching out eastward from the Helvellyn mass, and together with the very steep, rocky E face of the mountain they form the great hollow within which Red Tarn lies.

At the W end of Striding Edge, a scramble up a steep scree covered slope brings the climber to the flattish top of Helvellyn, and 300 m to the N lies the Triangulation Point, at 950 m (spot height 3118 ft) marking the summit. One may either return by way of Swirral Edge and the foot of Red Tarn, or descend the much smoother W slope across rough grazing land and through conifer plantations below 500 m (1700 ft contour) to the road alongside Thirlmere.

The features which have been described above are smaller in scale than the Swiss Alps considered in the preceding Study, but if prolonged warming of the Alpine climate occurred and the glaciers and snowfields melted, the landforms revealed would exhibit broadly similar characteristics.

The rotational and grinding movements of the ice in a névé basin together with a 'plucking' action which occurs when ice forces itself into crevices round a jutting piece of rock and eventually pulls it away, deepen the hollow and steepen its sides to produce a deep semi-circular embayment in the mountain such as that in which Red Tarn is situated. This feature of glaciated uplands is a **corrie** (also known as a **cwm** in Welsh, or a **cirque** in French). The existence of **corrie lake**, or tarn, in many corries suggests that the floor of the corrie forms a depression within which the lake water collects, and many corries do in fact, have a long section of this character.

In such cases there was some upward movement of ice as it was forced by pressure out of the névé basin. Note that such a basin could not be formed by water action since water cannot flow uphill. After the ice has melted the corrie fills with water until it overflows and this outflowing stream will often have deepened its outlet to form a notch in the rock barrier—thus gradually lowering the level of the lake surface.

When two or more névé hollows occur close to each other on the flanks of a mountain mass their enlargement during glaciation will result in the narrowing and steepening of the divides between them. In this way, the narrow ridges such as Striding Edge and Swirral Edge are formed—remnants of once broader interfluves. These features are **arêtes**. The arêtes were originally even steeper than they appear today. Now the lower slopes are mantled by great accumulations of scree which is still forming due to the frost shattering of the bare rock of the upper slope, but which must have formed much more rapidly in the immediate post-glacial epochs when temperatures were still sub-arctic, and a greater expanse of bare rock was exposed.

There are cases, such as the Matterhorn, where a mountain mass has been attacked on all sides by corrie forming action, leaving only a very steep rocky peak from which arêtes radiate. This is a **pyramidal peak.** But in the case of Helvellyn the corries all occur on the N and E facing slopes. This is quite a common occurrence in the Lake District and other British mountain areas. (Can you suggest why corrie formation was more active on north and east facing slopes?)

FURTHER WORK

Practical

(1) Use the photograph of Striding Edge as the basis for an annotated field sketch showing corries, arêtes, tarn, scree slopes.

(2) Draw a cross-section from Nethermost Cove to Red Tarn and on to Brown Cove; and a section E–W through the summit of Helvellyn to Red Tarn and along Red Tarn Beck. Annotate your sections to indicate significant physical features.

Reading

Carter: *Land Forms and Life*, Section 10, pp. 76–84
Dury: *The Face of the Earth*, Ch. 12, pp. 139–140
*Embleton: 'Snowdonia'. *British Landscape through Maps Series*, Geog. Ass.
Gresswell: *Glaciers and Glaciation*, Ch. 2, pp. 14–24
——: *Physical Geography*, Ch. 19, pp. 232–233; Ch. 23, pp. 313–314
Hardy & Monkhouse: *Physical Landscape in Pictures*, pp. 55–56
Holmes: *Principles of Physical Geology*, Ch. XX, pp. 641–650
Horrocks: *Physical Geography and Climatology*, Ch. 6, pp. 98–100
Monkhouse: *Principles of Physical Geography*, Ch. 8, pp. 189–190
——: *Landscape from the Air*, pp. 26, 27, 49
*——: 'English Lake District,' *British Landscape through Maps Series*. Geog. Ass.
Scovel: *Atlas of Landforms*, 106–109
*Sparks: *Geomorphology*, Ch. 12, pp. 268–274
Strahler: *Physical Geography*, Ch. 26, pp. 391–392; Ex. 1, pp. 407–8
Trueman: *Geology and Scenery* Ch. 14, pp. 185–202
Wooldridge & Morgan: *An Outline of Geomorphology*, Ch. XXII, pp. 344–347

Fig 114 Striding Edge, Swirral Edge and Red Tarn, Helvellyn

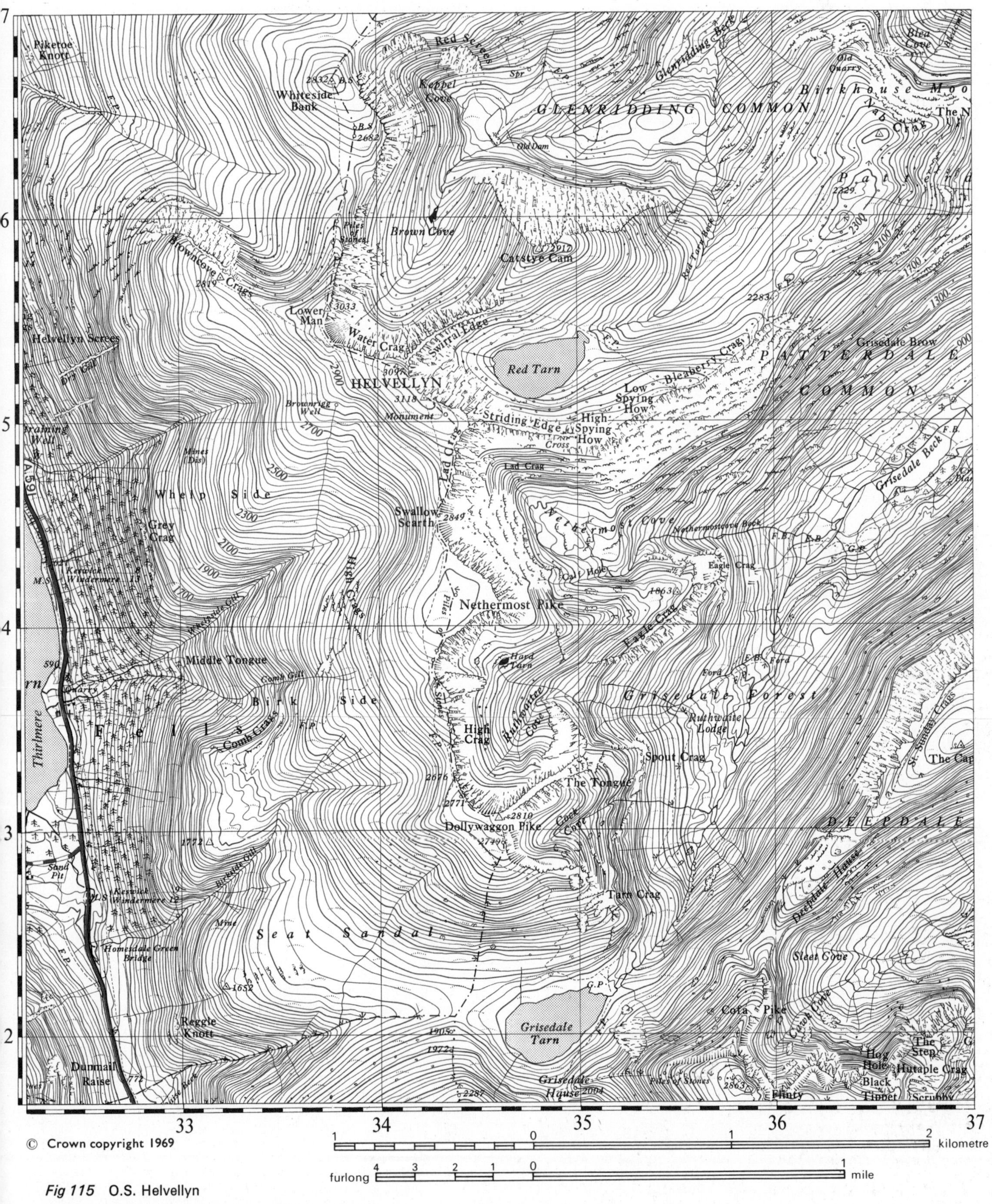

Fig 115 O.S. Helvellyn
Scale 1 : 25 000 (4 cm to 1 km *or* $2\frac{1}{2}$ in to 1 mile)
Contour interval 25 ft (7·5 m)

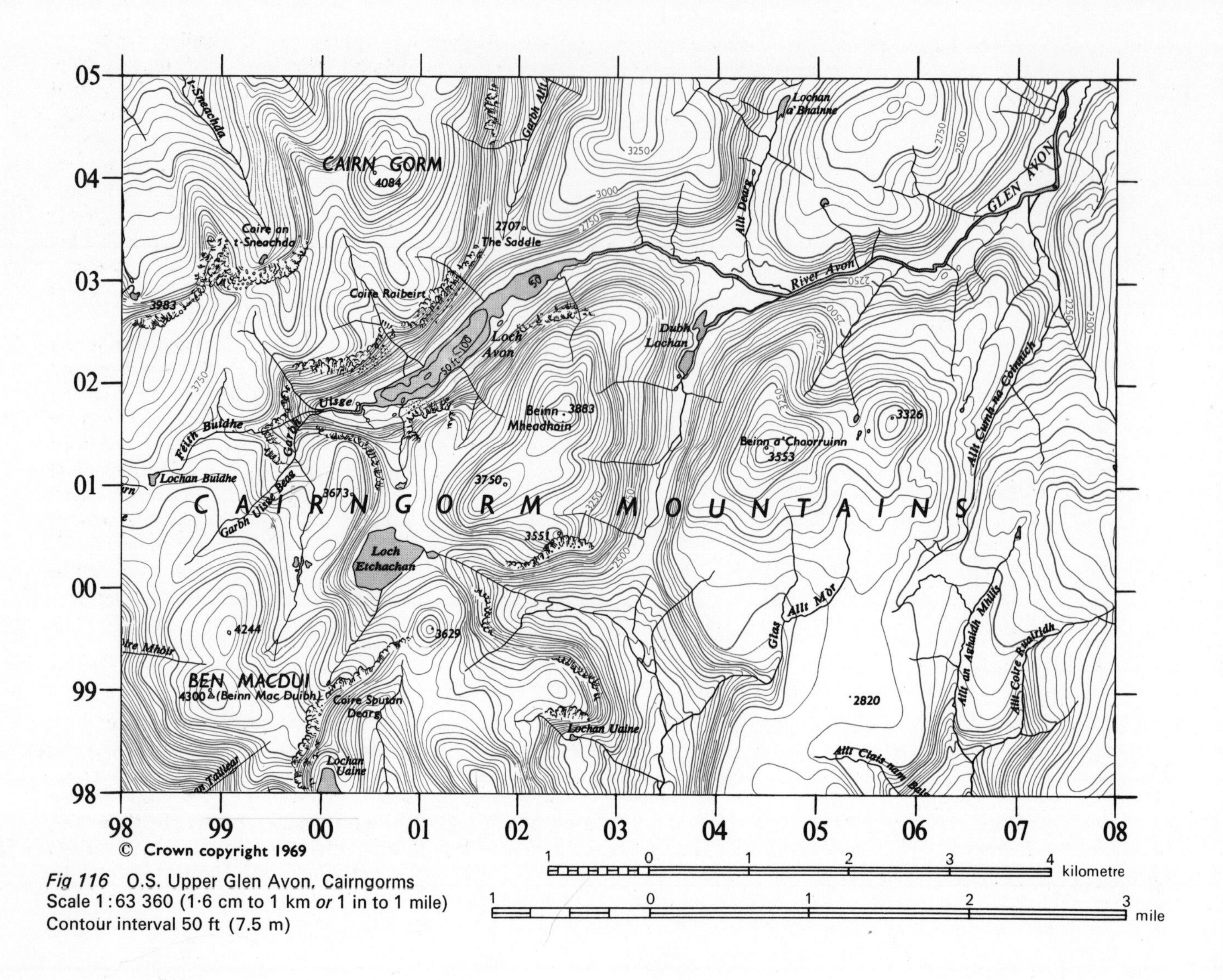

Fig 116 O.S. Upper Glen Avon, Cairngorms
Scale 1:63 360 (1·6 cm to 1 km *or* 1 in to 1 mile)
Contour interval 50 ft (7.5 m)

Upper Glen Avon, Cairngorms

Study C3

FEATURES OF A GLACIATED VALLEY,

O.S. 1:63 360 (1·6 cm to 1 km *or* 1 in to 1 mile)
Cairngorms (Tourist Edition) *or* Sheet 38 *or* 41

MAP ANALYSIS

(a) Draw a longitudinal profile of the Garbh Uisge Beag from 985003 to 040033. Note that the depths shown in Loch Avon are in feet below lake level, and these should be shown on your section.

(b) Draw a cross-section from Cairngorm (spot height 4084 ft) to Beinn Mheadhain (spot height 3883 ft). Where contours are exceptionally tightly packed it will not be necessary to plot every contour.

(c) Draw a much simplified sketch map of the main part of the valley. Locate the following features on *Fig. 116* and show their location by annotation on your sketch map.

(i) **Truncated spur**—this is a spur whose projection into the main valley has been removed in the process of valley straightening, leaving a very steep cliff-like slope overlooking the valley.

(ii) **Hanging valley**—a tributary valley which, in its upper section, appears adjusted to a level well above the main valley floor. Its stream then plunges over a very steep descent to enter the main valley.

(iii) **Rock basin**—a hollow in the longitudinal profile of the main valley, formed by the ability of ice to flow uphill under pressure. Such hollows are frequently filled by long narrow lakes—**ribbon lakes.**

(d) Attempt a field-sketch of the valley looking SW from 030030—even a very unartistic drawing can be useful in clarifying your ideas. One way to start is to use your cross-section as a rough guide.

(e) Summarize all the preceding work by writing an account of this valley, taking particular care to show how it differs from a valley formed entirely by river action.

The valley of Garbh Uisge Beag illustrates well the features of a glaciated valley of Highland Britain.

Perhaps the most generally recognized feature is the U-shape of the cross-section. As shown above (*Study C1*) the glacier would have largely filled the valley and the increased width and cutting power compared with the pre-existing stream results in the overdeepening of the valley and its conversion from a V- to a U-shaped section (*Fig 118*). It is sometimes possible, by drawing accurate cross-sections, to interpret the maximum height of the surface of the glacier with reference to the **shoulder** marking the top of the U-shaped valley.

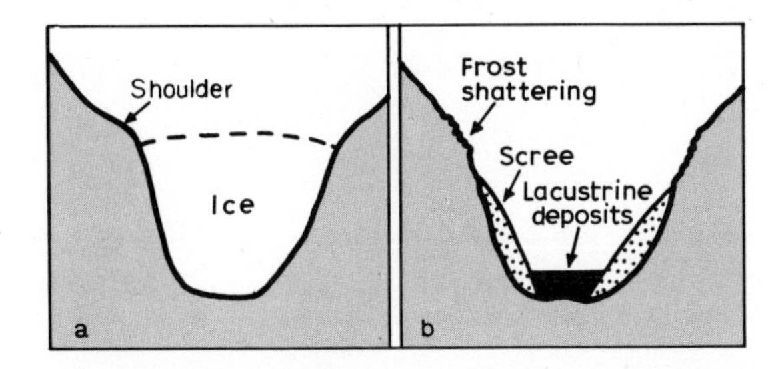

Fig 117 (a) U-shape valley cross section produced by glacial erosion (b) Post glacial modification of valley cross section

Although many valleys exist with the idealized smooth U cross-section, it is equally common to find that post-glacial conditions have resulted in major modifications. The previous Study mentioned that in immediately post-glacial times, when minimum temperatures were still very low, considerable frost shattering occurred on valley sides particularly at upper levels. This still occurs today though on a much reduced scale. The rock fragments thus dislodged fell by gravity and accumulated to form scree slopes which obscure and protect the lower portion of the glacial trough. Secondly, long narrow lakes often form in glaciated valleys as a result of morainic damming, of temporary ice damming or the occurrence of rock basins. These gradually silt up, and are finally drained by river erosion lowering their outlets thus exposing a flat valley floor of alluvium over which meanders a small present-day stream clearly having no relevance to the erosion of the large valley—hence the term **misfit stream.**

In its long profile, map analysis has shown that the Garbh Uisge Beag drops steeply into the overdeepened trough at 998016. This **trough end** form at the head of glaciated valleys occurs where a number of minor glaciers converged from their corries. The resultant increase in volume led to a marked increase in erosional power, and thus to the sculpting of the overdeepened valley below this point. The longitudinal profile demonstrates clearly the ability of a glacier to hollow out rock basins along its course. These may be related either to reduced rock resistance or increase in glacial volume, perhaps due to the addition of ice from large tributary glaciers, or to a combination of these factors.

The smooth course and parallelism of the 2500 ft contour (750 m) on both sides of the valley provides striking evidence of the way in which glaciers remove obstacles from their path. If this is compared with the much more sinuous course of the 3500 ft contour (1050 m), especially on the slopes of Cairngorm, it can be seen how spurs have had their lower portions removed—**truncated.** Finally, a study of the contour pattern of valleys such as Coire (Gaelic equivalent to 'corrie') Raibeirt, or of the stream rising at 008010, particularly when considered in relation to a longitudinal profile of these streams, shows the features of hanging valleys.

In the two preceding Studies, those features generally associated with upland glaciation have been analysed. No mention has been made of the beauty and grandeur of such landforms. This is essentially landscape which must be travelled on foot, and the student who has not walked in Snowdonia, or the Lake District or the Scottish Highlands must always realize that maps and photographs are but a poor substitute.

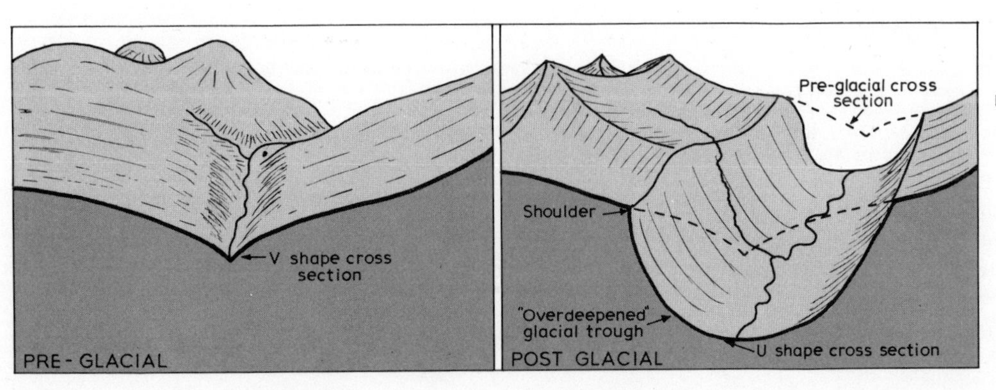

Fig 118 Idealized diagram to show modification of valley cross section by glacier

FURTHER WORK

Practical

(1) As suggested in the last paragraph of the above chapter, every student of the landscape should take any opportunity offered to visit a glaciated region. Even if such a visit is not an organized field study, any student should return with a mass of field sketches, photographs for annotation, and notes and questions on features seen.

(2) The techniques of map and photograph analysis as suggested in the preceding studies may be usefully applied to other glaciated areas. O.S. 1:63 360 sheets especially suitable are:

Lake District Tourist Edition *or* Sheets 82 and 83
Sheet 107, Snowdonia
Sheet 46, Loch Linnhe
Lorn and Lochaber Tourist Edition

(3) See also task 2 following *Study E8.*

Reading

Dury: *The Face of the Earth*, Ch. 12, pp. 140–147
Gresswell: *Glaciers and Glaciation*, Ch. 3, pp. 25–51
——: *Physical Geography*, Ch. 23, pp. 297–309
Hardy & Monkhouse: *Physical Landscape in Pictures*, p. 57
Holmes: *Principles of Physical Geology*, Ch. XX, pp. 650–655
Horrocks: *Physical Geography and Climatology*, Ch. 6, pp. 100–104
Monkhouse: *Principles of Physical Geography*, Ch. 8, pp. 191–193
——: *Landscape from the Air*, pp. 28, 48
Scovel: *Atlas of Landforms*, pp. 112–113
*Sparks: *Geomorphology*, Ch. 12, pp. 274–277
Strahler: *Physical Geography*, Ch. 26, pp. 392–393
Trueman: *Geology and Scenery* Ch. 15, pp. 203–208
Wooldridge & Morgan: *An Outline of Geomorphology*, Ch. XXII, pp. 337–344

GLACIAL DEPOSITION

Study C4

Over the greater part of Lowland Britain N of a line from Bristol to London, many of the minor landscape features, and even some of the major ones, are attributable to the Pleistocene Ice Sheets. As ice sheets engulfed the British Isles from the N and from across the North and Irish Seas, and then haltingly receded, they brought with them an enormous mass of debris eroded from our own highland areas and from the Scandinavian mountains. Towards their southernmost limits the ice sheets lost the impetus provided by the snow fields of still colder regions; their energy was reduced so that the ice was now overloaded with debris and as the ice slowly melted this great agent of erosion became an agent of deposition. Any detailed analysis of these epochs must be extremely complex since deposits of one era were subject to fluvial erosion before being overridden by the ice of the next glacial advance—a sequence which might be repeated many times. As a result of this it is rarely possible to attempt a general interpretation of a landscape of glacial deposition until very precise observations from a wide area can be correlated. Nevertheless isolated features can be recognized in the field, and from photographs, and some of these are studied in this section.

Boulder-clay, sometimes alternatively referred to as **till**, mantles large areas of Lowland Britain, as well as forming extensive sheets in Northern Europe and North America. Just as any surplus load in a river is deposited to form alluvium, particularly when river energy declines, so the load of a glacier or ice sheet is deposited towards its margins as **ground moraine.** This mainly consists of finely ground particles of rock—so fine as to be almost a clay in character. When an ice sheet is in retreat, however, the ice at its margins becomes virtually stagnant, and as this melts all the load contained within it is left in a confused jumble of material with sizeable stones and boulders, sandy and clay materials all mixed together. The precise composition and nature of this boulder-clay depends on the nature of the rocks from which it is eroded, and on the degree of grinding down which it has undergone during transport. Thus the predominantly chalky boulder-clay of parts of East Anglia is very different from the stony boulder-clay fringing some of our mountain areas. In the field, it is the completely unstratified nature of the deposit as exposed in cuttings, quarries or cliffs which gives the clue to its origin (*Fig 119*).

The terms boulder-clay and till are often used to include the ground moraine which was deposited beneath the ice whilst it was still in motion. Just as moving water leaves ripple marks in sand, one may envisage the moving ice moulding the ground moraine into an undulating surface. These undulations characteristically take the form of **drumlins**. The photograph (*Fig 120*) shows a low-undulating landscape with the tops of the rounded hills about 15 m above the lowest ground. Each of the hills is roughly oval in plan, though with one end frequently wider than the other, and with the highest point situated at the broader end of the hill. These drumlins and the intervening lowland are of boulder-clay and the depressions are frequently badly drained and even marshy. The long axes of the drumlins are parallel to each other and all have their broader end facing in the same direction. Such a shape clearly suggests some form of streamlining—comparison with, for example, a whale, will suggest that the broad end faces towards the direction from which movement comes. Since it has been shown that the drumlins were formed beneath moving ice, it is easy to see that the wider, higher end of the hill indicates the direction from which ice moved, parallel to the long axes of the hills. This conclusion can be substantiated by reference to other data on the movement of ice sheets.

One such source of information on the direction of ice sheet movement is provided by the occurrence of recognizable rocks either in the boulder-clay or remaining as isolated boulders (**glacial erratics**) on the surface. The existence in the boulder-clay of Essex of chalk from the East Anglian Heights, Coal Measures, dolerite and Mountsorrel Granite from the East Midlands, Red Chalk from Lincolnshire and even rocks of Scandinavian origin enable one to determine the origins of the ice which deposited this material. (*Figs 119 and 121.*)

Fig 119 Boulder clay

Terminal moraines—At the close of the analysis of the Rhône Glacier (*Study C1*) reference was made to the accumulation of glacial debris at the snout of the glacier to form a terminal moraine. Such features indicate positions at which the ice front has remained stationary for considerable periods—the size of the feature being dependent on glacial load and the time during which the ice front occupied the site. Terminal moraines form lines of low hills predominantly stony or sandy in

Fig 120 Drumlin Topography, County Down, N. Ireland

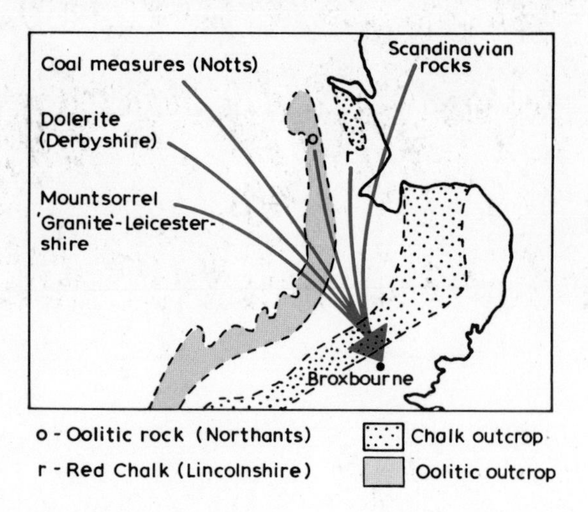

Fig 121 Origins of material in Boulder Clay of North Essex (after Dury)

character, since finer material will have been removed by melt waters; and subsequent weathering may lead to the formation of a thin veneer of soil.

The preceding features were a direct consequence of glacial deposition. But towards the margins of an ice sheet, melt water within the glacier is increasing in volume and a very considerable flow of water emerges from the ice front. As such streams emerge from caves in the ice front they are normally flowing with a high velocity and are already heavily loaded with debris collected from the ice during the water's passage through crevasses and ice-caverns. In addition they collect and transport morainic material already deposited by the ice along its margins, or laid down below the current ice front when the ice sheet was less extensive. The term **glaci-fluvial deposits** is applied to all the debris of such streams indicating that the material of which they are composed was glacial in origin but that running water was the final agent of transport and deposition. All glaci-fluvial deposits can be distinguished from boulder-clay by the existence of some degree of stratification. The heavy load of melt water streams, combined in many cases with a low gradient away from the ice front, results in deposition—the streams become braided and as their velocity decreases the nature of deposits becomes progressively finer, ranging from coarse gravel close to the ice front through sand, and fine sand to clays as the distance from the ice increases.

Where a glacier or ice sheet retreats from a terminal moraine, the latter may form a dam ponding back melt water and forming a temporary lake. Into this lake will pour the glacial streams, and their debris sinking through the still waters accumulates as **lacustrine sediments**, i.e. deposited on the bed of a lake. Since the supply and nature of debris varies with the seasonal temperature regimes, the lacustrine deposits are often arranged in layers of coarse and fine material. Each layer of coarser sediment represents the increased summer flow of melt water, and from these layers, or **varves** glaciologists have been able to date events of past glaciations.

At the point where a melt water stream enters such a lake, its velocity is suddenly checked and a delta is formed—a **kame**. The sinuous course of a melt water stream beneath the ice is sometimes preserved by its sandy deposits as an **esker** (*Fig 123*)—a winding ridge reminiscent in size and form to a railway embankment—which curves its way over a glaciated lowland without any relationship to the geology of the area.

The following pair of block diagrams illustrate the relationship between the features described in the preceding section to the position of the ice sheet and to each other. *Fig 122a* shows conditions during the glacial epoch and *Fig 122b* shows the features which remain after the ice has melted.

FURTHER WORK

Practical

(1) If you live in, or can visit, an area where boulder-clay occurs, use a Drift Edition of the Geological Map to locate it, and try to find an exposure of the deposit—e.g. in a quarry, on a cliff, at a building site, etc. Study the

Fig 122 Marginal landforms of continental glaciers
(a) With the ice front stabilized and the ice in a wasting, stagnant condition, various depositional features are built by meltwaters
(b) After the ice has wasted completely away, a variety of new landforms made under the ice is exposed to view

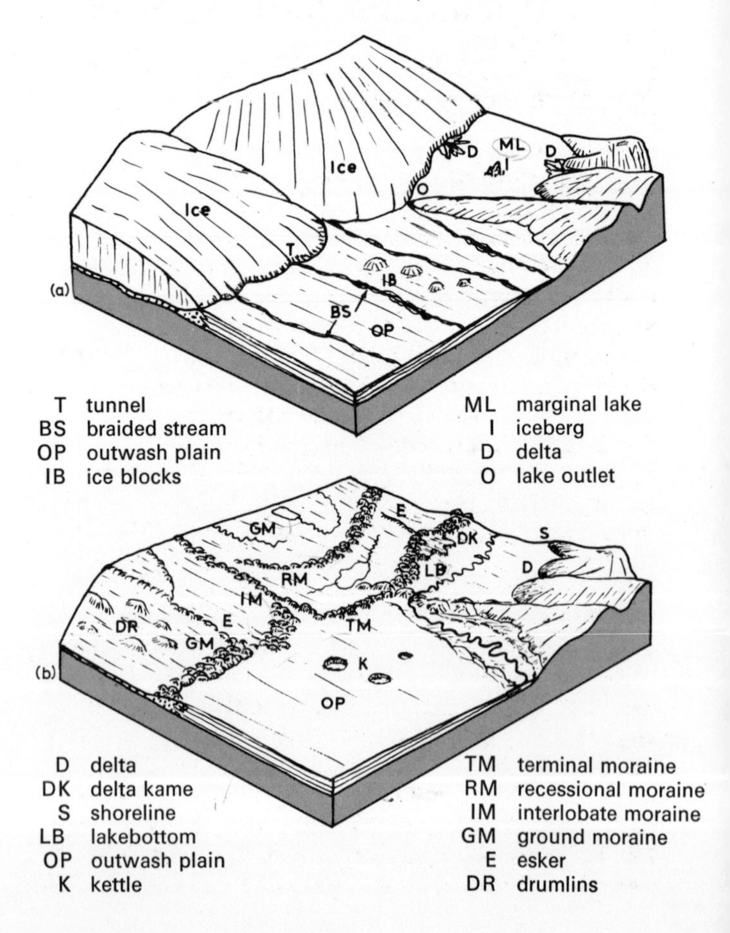

material of which the clay is composed to see whether its origin can be determined. Collect and identify specimens of stones in the boulder clay and try to find where they may have come from.

(2) Draw simple diagrams to illustrate the mode of formation of each feature of glacial deposition described above and write two or three sentences to summarize the characteristics and origin of each feature.

Fig 123 The Punkaharju esker, Finland

Reading

Anderson: *Splendour of Earth*, pp. 332–338
Carter: *Land Forms and Life*, Section 11, pp. 85–89
Dury: *The Face of the Earth*, Ch. 13, pp. 154–160
Gresswell: *Glaciers and Glaciation*, Ch. 4, pp. 52–70; Ch. 5, pp. 71–86
——: *Physical Geography*, Ch. 20, pp. 241–260
Hardy & Monkhouse: *Physical Landscape in Pictures*, pp. 58–59
Holmes: *Principles of Physical Geology*, Ch. XX, pp. 658–668
Horrocks: *Physical Geography and Climatology*, Ch. 6, pp. 109–113
Monkhouse: *Principles of Physical Geography*, Ch. 8, pp. 195–203
——: *Landscape from the Air*, pp 30, 31
Scovel: *Atlas of Landforms*, pp. 114–127
*Sparks: *Geomorphology*, Ch. 13, pp. 288–298
Strahler: *Physical Geography*, Ch. 26, pp. 400–406; Exs. 3 and 4, pp. 410 and 411
Trueman: *Geology and Scenery*, Ch. 9, pp. 127–129; Ch. 10, pp. 140—141; Ch. 13, pp. 176–177
*Wooldridge & Morgan, *An Outline of Geomorphology*, Ch. XXII, pp. 347–358

North Yorkshire Moors

Study C5

DRAINAGE MODIFICATIONS DUE TO GLACIATION

O.S. 1:63 360 (1·6 cm to 1 km *or* 1 in to 1 mile) Sheets 86 Redcar and Whitby; 92 Pickering; 93 Scarborough
or O.S. 1:63 360 (Tourist Edition), North Yorkshire Moors

On an atlas map, study the course of the R. Derwent which rises NW of Scarborough. This stream first flows SE off the Moors towards the coast, but then turns S to parallel the coast before swinging away inland and flowing SSW to join the Ouse, so that its waters eventually reach the sea in the Humber some 65 km S of Scarborough. The 1:63 360 map shows the even more surprising fact that there are clear cut valleys by which the river could reach the sea N of Scarborough, but in each case these are spurned and the river turns S from them, cutting deep, narrow gorges through high ground. Similarly the Derwent leaves the Vale of Pickering by cutting the Kirkham Abbey Gorge S of Malton, ignoring the apparently easier outlet eastward through low hills to Filey.

It is important that the relative positions of these features and of the areas considered in detail below should be studied from an atlas or other suitable map.

Next consider the feature of Newton Dale, *Figs 126, and 128*. The dramatic character of this valley can be appreciated if a cross-section is drawn from 840940 to 825955, and the section compared with the photograph which shows the same portion of the valley. One sees a remarkably sharply defined valley 75 m deep (i.e. 500–750 ft contours) with very steep sides cut into the plateau-like surface of the moors. An even more surprising characteristic is that Newton Dale cuts its course right across the divide between N and S flowing streams. The highest point along the valley floor, 165 m (spot height, 550 ft) is reached at about 850975 and is marked by an area of marsh and indeterminate drainage; certainly there is no clear valley head.

One may next consider the area which lies to the south of the upper Esk Valley (*Fig 124*). Five fairly flat floored dales drain north to join the Esk (Westerdale, Danby Dale, Little and Great Fryup Dales and Glaisdale). Some features of significance may be observed if a careful study is made of the contour patterns of the dividing ridges. Between Great and Little Fryup Dales, at 719049 there is a clearly marked break in the dividing ridge, and close examination reveals similar though less obvious nicks in dividing ridges at 683049, 681072, 766050 and 753067. In the field, these features are more apparent than on the map, and in form suggest channels eroded across the spurs, similar in cross section to Newtondale, though of course much smaller.

To the data which have been indicated above, can be added other evidence dependent upon field observation:

(a) The floors of the dales draining to the Esk, and also the area of the Vale of Pickering can be shown to be formed of **lacustrine deposits** (i.e. deposits from the bed of a lake);

(b) The land enclosed by the 100 ft contour (30 m) on which is the site of the town of Pickering (799838) is deltaic in origin;

(c) The low hills along the coast between Scarborough

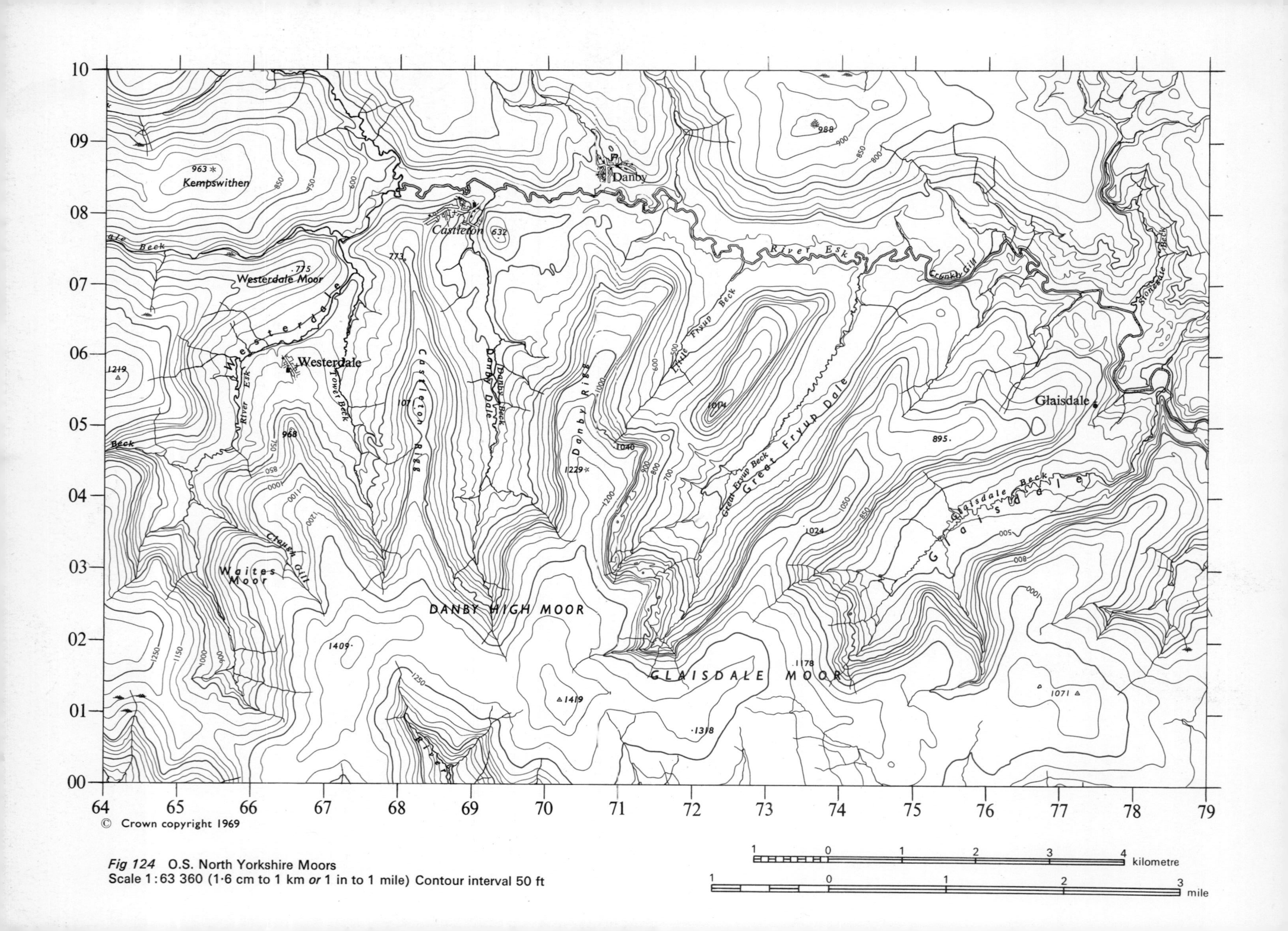

© Crown copyright 1969

Fig 124 O.S. North Yorkshire Moors
Scale 1:63 360 (1·6 cm to 1 km *or* 1 in to 1 mile) Contour interval 50 ft

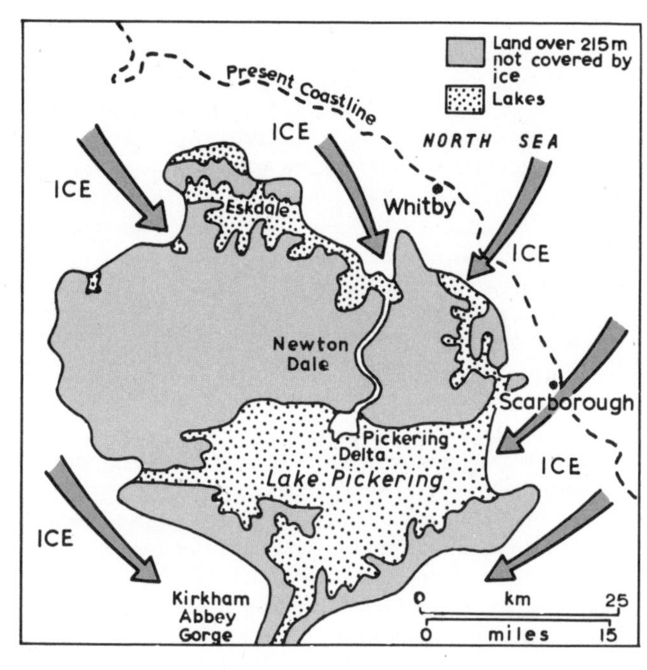

Fig 124a North York Moors at one stage of ice advance (after Kendrew)

and Filey are predominantly of boulder-clay.

All this evidence leads to the conclusion that the contemporary landscape cannot be explained in terms of the normal stages of river development, nor by reference to the types of glacial activity previously considered. The district in fact provides an excellent example of the way in which drainage patterns may be temporarily or permanently changed by the presence of ice sheets.

Before explaining the sequence of events which are believed to have occurred in this region, it must first be appreciated that it is possible for an ice sheet, or glacier, to exist without all water in the area being permanently frozen. One example which may be cited to illustrate this point is the Marjelen See which is a lake in the tributary valley to the Aletsch Glacier, Switzerland (*Fig 125*). The lake owes its existence to the fact that water is dammed back by the glacier blocking the main valley. The Franz Joseph Glacier, South Island, New Zealand, extends well below the tree line to move through a forested valley, illustrating again that ice masses can extend well beyond their place of origin into climates of less intense cold.

During the last glaciation the ice sheets, up to 300 m thick in places, spread southward slowly and surrounded the North Yorkshire Moors area, leaving it as an 'island' of uncovered land (*Fig 124a*). The first effect was to block the mouth of the River Esk and prevent its waters reaching the sea. At the same time ice advancing along the western edge of the moors blocked the possible alternative outlet via Kildale (605095). Gradually water

Fig 125 The Marjelensee, Switzerland

accumulated in the dales forming lakes which at various times found temporary outlets across the spurs from one dale lake to another—hence the cols or **spillways** observed above. Eventually these lakes amalgamated and the water surface stood 215 m above present sea-level, coincident approximately with the 750 ft contour. At this height the water reached the level of the lowest point in the main divide and began to overflow southwards in an increasing torrent which eroded the trench of Newtondale known as a **glacial overflow channel**. By this time both the E and W ends of the Vale of Pickering were blocked by southward extensions of ice so that another vast lake was formed. The Newtondale overflow entering this lake deposited the Pickering Delta.

Ice advancing down the east side of the moors and extending out to sea far beyond the present coastline had meanwhile blocked the mouth of the upper Derwent, which found an alternative outlet southwards, cutting in the process the deep valleys of Langdale (936935) and Forge Valley (985865), thus adding to the waters of Lake Pickering. This lake, like the others described above, rose in level until at a height of about 75 m above present sea-level—an altitude indicated by the 250 ft contour—it found and deepened an outlet at the Kirkham Abbey Gorge (732660).

From this point on, the story is of the gradual retreat of ice. Along the E coast S of Scarborough, boulder clay masses remained to obstruct the eastern end of the Vale of Pickering, and the upper Derwent maintains its ice-deflected course to the present day, crossing the floor of the long since drained Lake Pickering. To the N of the Moors however, the withdrawal of ice allowed the R. Esk to return to almost its pre-glacial course, abandoning the outlet via Newtondale. As the ice level receded various temporary links between the lakes were established either deepening notches in the dividing spurs (*Fig 127a*) or cutting terraces between the hillside and the ice front (*Fig 127b*).

In this study we have first observed a number of features in an area which seem to require special explanation and have traced the events, spread over many thousands of years, which may be invoked to explain these features. The student should be aware of two further points: firstly the elucidation of the stages of development here described can only be the result of long and painstaking field research. Map study can, and should, raise questions but it is the detailed analysis of landscape features which must provide the answers.

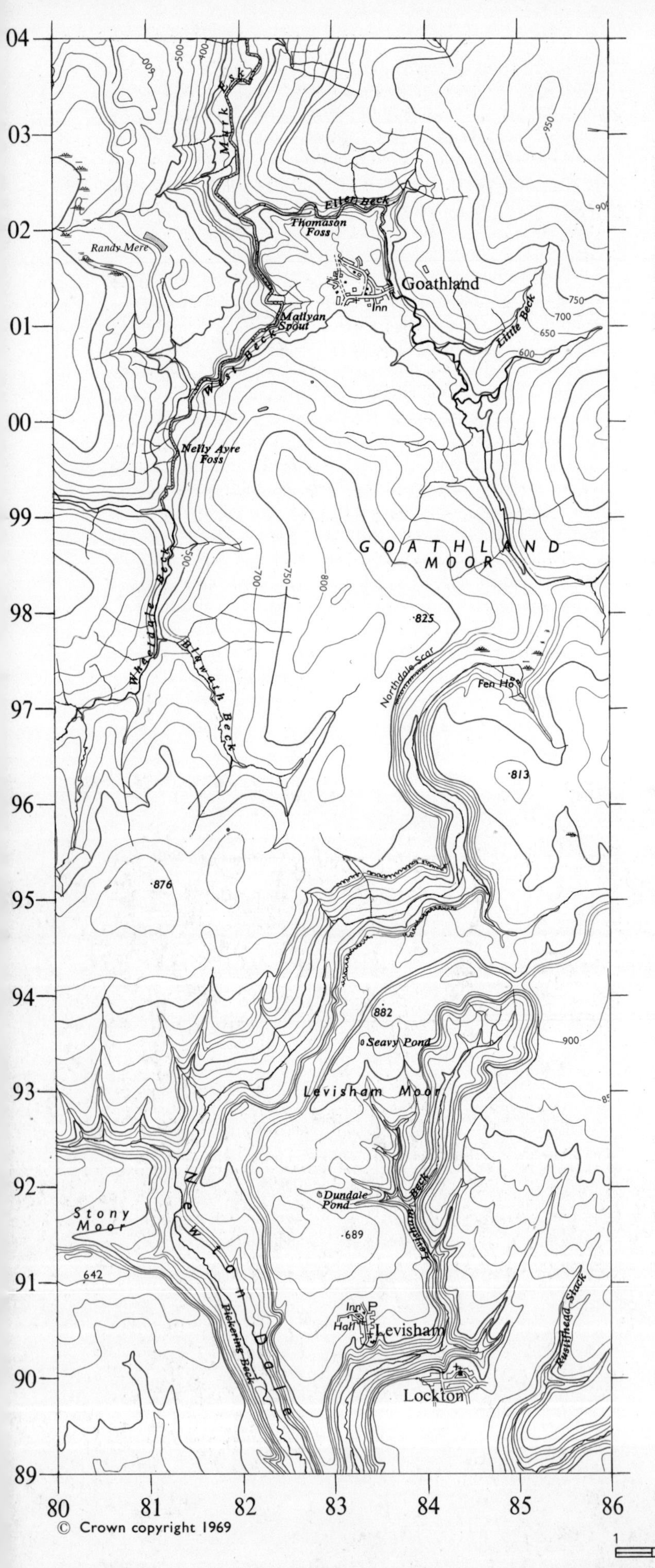

Fig 126 O.S. Newtondale
Scale 1:63 360 (1·6 cm. to 1 km *or* 1 in to 1 mile)
Contour interval 50 ft

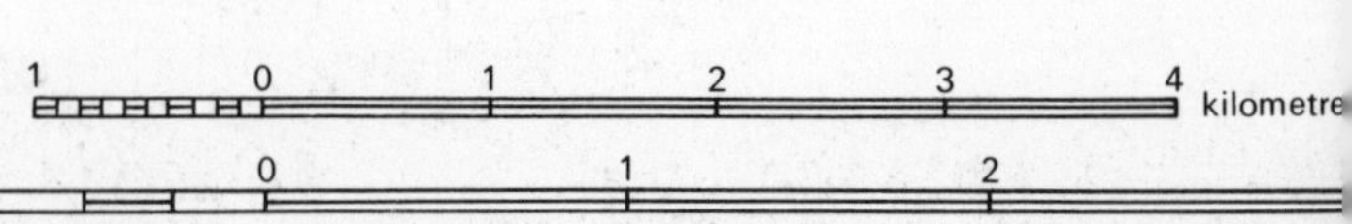

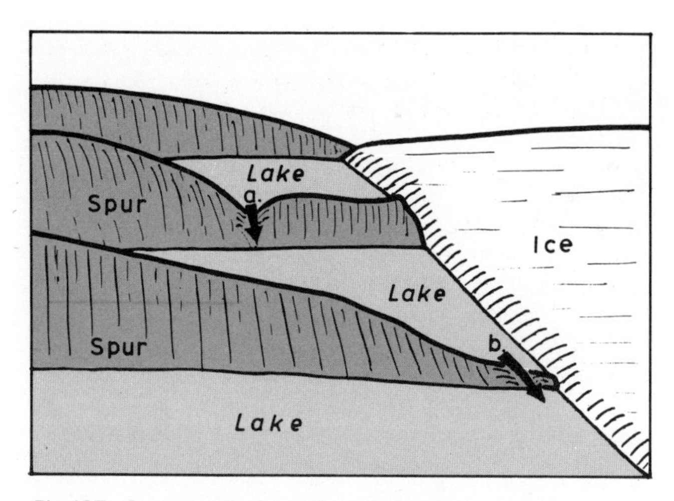

Fig 127 Some types of overflow channels

Secondly, the student should realize that the stages in the evolution of features described in this study are, in fact, more complex and have been worked out in much more detail than it is possible to demonstrate here.

Many other examples exist to show that glaciation has altered the pattern of drainage which existed before the invasion of an area by ice. Ice advancing southward across the Cheshire Plain ponded up the waters of the upper Severn which had previously flowed N to join the R. Dee. Water from this lake (Lake Lapworth) escaped through and eroded the Iron Bridge Gorge, and the modern course of the R. Severn perpetuates this diversion.

During the penultimate glaciation a vast lake filled much of the Midland Plain of England (Lake Harrison). Prior to glaciation the area was drained by a much longer R. Soar, rising near Bredon Hill; Lake Harrison, however, was finally drained towards the SW so that the area today lies mainly in the basin of the R. Avon draining to the Severn.

Prior to the Pleistocene Glaciation at a time when sea-level was some 120 m higher than at present the R. Thames followed a much more northerly course across the London Basin passing close to the present sites of Watford, St. Albans and Ware. Its diversion to the present course has been shown to be due to the invasion of the northern London Basin by ice at the maximum of its advance.

Fig 128 Newtondale under snow

The geological time-scale (*Fig 1*) shows that the numerous advances and retreats of the Pleistocene (Quaternary) Ice Sheets occurred 'as if 'twere yesterday' in the history of our landscape. *Fig 110, page 74* showing these fluctuations in greater detail, makes it clear that the period separating us from the last ice advance is considerably shorter than the longest inter-glacial epoch. We could therefore be in an inter-glacial period at present so that it is hardly surprising to find that many of the familiar features of our landscape—lowland as well as highland—can in part be attributed to glacial influence. No study of landscape evolution in Britain can be complete which does not take into account the possible erosional and depositional effects of the Pleistocene Ice Age.

FURTHER WORK

Practical

(1) Use data obtained from some of the reading suggested below to explain glacial modifications in drainage in one of the areas mentioned above.

(2) Can you suggest what the landscape of your home area would be like if there had been no Pleistocene Ice Age?

Reading

Dury: *The Face of the Earth*, Ch. 14, pp. 164–178; Ch. 17, pp. 215–216

Gresswell: *Glaciers and Glaciation*, Ch. 6, pp. 87–96

——: *Physical Geography*, Ch. 21, pp. 261–274

Holmes: *Principles of Physical Geology*, Ch. XX, pp. 668–671, *683–688, 709 (fig)

Horrocks: *Physical Geography and Climatology*, Ch. 6, pp. 104–109

Monkhouse: *Principles of Physical Geography*, Ch. 7, pp. 168 and 171–172

——: *Landscape from the Air*, p. 32

*Sparks: *Geomorphology*, Ch. 13, pp. 298–309; Ch. 12, pp. 280–284

Trueman: *Geology and Scenery* , Ch. 3 pp. 35–42

*Wooldridge & Morgan: *An Outline of Geomorphology*, Ch. XXII, pp. 358–369

Loch Linnhe, Scotland **Study C6**

A FIORD COAST AND THE INFLUENCE OF FAULT SYSTEMS

O.S. 1 : 63 360 (1·6 cm to 1 km *or* 1 in to 1 mile)
Sheet 46, Loch Linnhe
or O.S. 1 : 63 360 (Tourist Edition), Ben Nevis and Glencoe

MAP EXERCISE

Note: Questions in brackets are intended to be worked in addition if the complete O.S. sheets are available.

(1) What map evidence on a fully coloured O.S. sheet would show that Loch Linnhe and Loch Leven are tidal —i.e. are sea-lochs, not fresh-water lakes. [Locate other large lochs which are fresh-water lakes.]

(2) Draw a cross-section from Beinn na Gucaig (063654) to Beinn na Cille (010672). N.B. Indicate the surface of the loch and show the underwater profile down to 10 fathoms, but remember that O.S. maps do not indicate submarine contours deeper than this. [Study the depth contours for Loch Shiel (given in feet) and compare the maximum depths with the altitude of the lake surface.]

(3) Draw a sketch map of the area shown on the extract; lightly shade the land below 50 ft (15 m) [and indicate by dots the location of all settlements and isolated houses].

(4) What evidence do you find that deposition of alluvium is occurring in the waters of the lochs? How might you explain the narrowing of the lochs at Corran Narrows (018635) and Ballachulish Ferry (053597)? [Use the submarine contours to sketch long profiles of Loch Etive and Loch Creran—indicate the position of places where these lochs narrow.]

(5) What map evidence suggests that the area shown on the map extract [or on the O.S. sheet] has been glaciated?

An atlas shows that long narrow inlets of the sea such as Loch Linnhe are characteristic of the W coast of Scotland, and that very similar characteristics may be seen in certain other coastal regions backed by glaciated uplands—e.g. British Columbia; Norway; the SW of South Island, New Zealand. Other characteristics common to all these **fiord** coasts are:

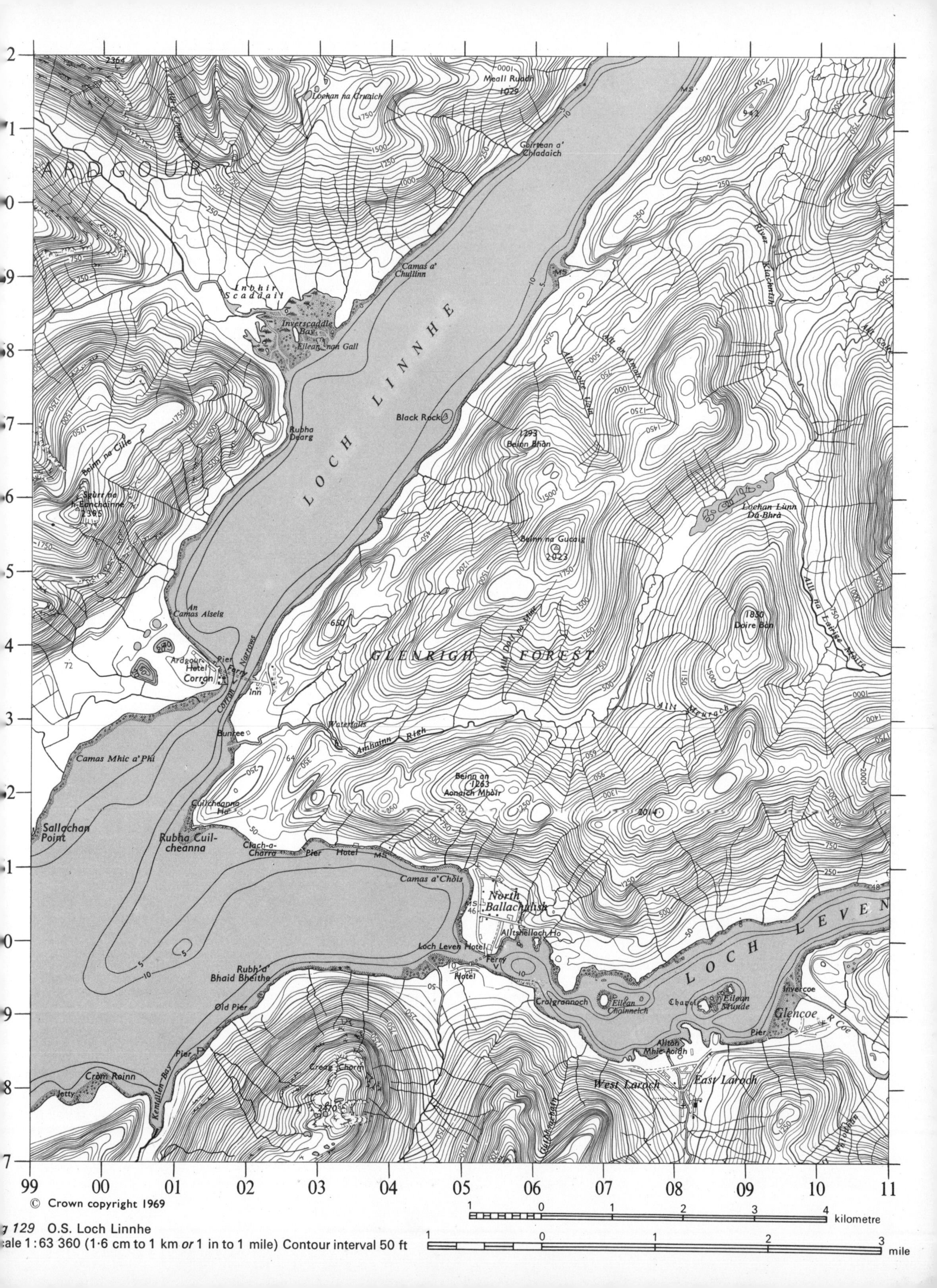

g 129 O.S. Loch Linnhe
ale 1 : 63 360 (1·6 cm to 1 km *or* 1 in to 1 mile) Contour interval 50 ft

(a) the relative straightness of sections of the fiords, unbroken by any major spurs;

(b) mountain sides which fall steeply to a very narrow shoreline;

(c) deep water reaching close to the shore, so that a U-shaped cross-section, continuous above and below sea-level, is apparent;

(d) occasional small areas of recent alluvium providing almost the only low-lying flat land in the area;

(e) points at which the fiord shallows and narrows, sometimes to such an extent as to separate a fresh-water lake from the sea water of the fiord.

Comparisons with the evidence discussed in *Studies C1, C2* and *C3* indicate that the landscape of the Loch Linnhe area has been affected by glaciation during the Pleistocene (Quaternary) Ice Age. A glacial origin of the lochs themselves is particularly suggested by their U-shaped cross-section and the evidence of the sculpting of rock basins in the long profile of, for example, Loch Leven. How does it come about that these glaciated valleys are now flooded by the sea? In the first place it must be remembered that ice erodes by virtue of its weight and volume and, since it is not limited in its vertical erosion by any base level, is able to carve out valleys below sea-level. Secondly, during a period of major ice advance, the normal cycle by which rivers return rain water to the sea is interrupted and sea level is lowered; twenty thousand years ago, for instance, during the last major ice advance, sea level was lower than at present (figures of between 60 m and 225 m lower have been suggested). Glacial valleys carved under such conditions would be flooded by the gradual rise of sea-level as the ice retreated, producing a deeply indented coast-line with many islands. Combined with such oscillations of sea-level the land itself has, to a lesser extent, been depressed by the weight of ice upon it and, as this load disappeared, the land has risen again by the process known as **isostatic recovery.** This latter process is frequently more delayed than the rise in sea level (a **eustatic** change) so that in Western Scotland there has been first the flooding of glaciated valleys to form fiords, followed by a rise in the land itself producing a **raised beach** approximately 7 m above sea-level. This narrow shelf has had a marked influence upon the location of coastal settlements and communications along the shores of the fiords. A further development since the disappearance of ice has been the construction of alluvial deposits, often in deltaic form, where sizeable streams bring down eroded material from higher up their valleys. Such features may be recognized at 080585, 030680, 065690, etc. Where shallower parts of the fiord occur between rock basins scooped out by the glaciers, a combination of isostatic uplift and post-glacial deposition may produce marked narrowings as at Corran Narrows (018635) and Ballachulish (053597). The Ordnance Survey map indicates vehicle ferries and a concentration of settlement at these points, emphasizing their importance in the human geography of the region.

Figure 130 is a sketch map of the area covered by O.S. 1:63 360 Sheet 46, showing the coastline and selected portions of rivers in the area. Since this map shows only

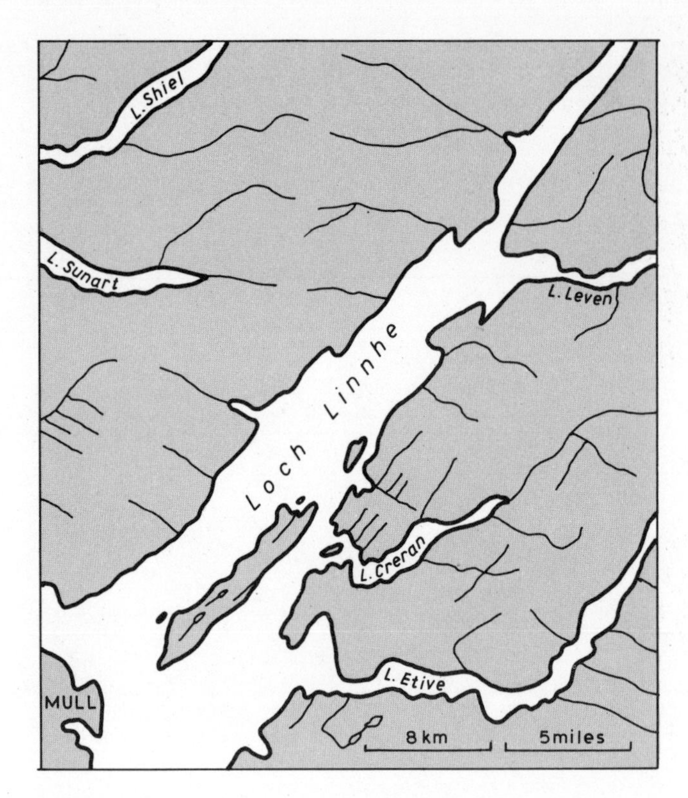

Fig 130 Drainage pattern of O.S. Sheet 46, showing influence of (a) SW-NE and (b) WNW-ESE structural trends

some of the streams it can have no value as conclusive evidence, but the student will notice the predominance of two directional trends. First there is the approximately SW–NE lines as shown in the coast of Loch Linnhe, of upper Loch Etive, parts of Loch Creran and many large and small valleys. Secondly there are the numerous lochs and valleys showing a WNW–ESE trend, e.g. lower Loch Etive, Loch Sunart and Glen Tarbert, etc. The existence of these trends can be verified from the O.S. map. It must clearly be of significance that these same directions can be recognized on the geological maps of the area, indicated in the directions of **dykes**[1] and faults, in particular the line of the Great Glen Fault which marks approximately the north shore of Loch Linnhe and continues NE through the Great Glen (Glen More) to Inverness. **Trend lines** such as these are often reflected in the drainage pattern of a region: faults are accompanied by zones of shattered rocks which are relatively less resistant. Rivers develop along these lines, and in the present instance valley glaciers followed more or less the pre-existing rivers thus emphasizing the structural trends.

In a number of preceding Studies (*A11, A12, A13, B5, B6, B7* and *B8*) passing reference has been made to vertical movements in the earth's crust, faulting of rocks, and other structural changes. Structural movements are the consequence of internal forces within the earth, so that the mechanisms involved are primarily the concern

[1] Vertical bands of intrusive igneous rock. (See p. 135)

of geophysicists. Their studies, based largely in the interpretation of the behaviour of earthquake shock waves, show that the earth has a thin **crust** underlain by a thick **mantle** enclosing a central **core**. *Fig 131* shows that density increases towards the centre and that the crustal layer is thin and is not continuous. More is known about the surface of the moon than of the chemical and physical nature of the centre of our own planet, but we are able to say that, with certain localized exceptions, the upper

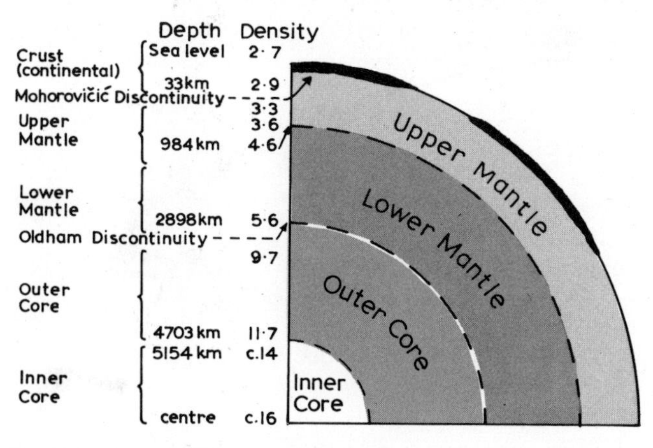

Fig 131 Internal structure of the earth

parts of the crust beneath the continents are relatively rich in **si**lica (65–75%) and that **al**umina is the next most common constituent. These rocks are referred to in abbreviated form as **SIAL**. As depth increases, the rocks of the lower crust become progressively more basic with less silica (c. 50%) and with iron and **ma**gnesia the next most important constituents. These rocks are known as **SIMA**.

Investigations show that sial does not occur beneath the deep ocean floor but is confined to the continental masses, and further that the crustal rocks of relatively low density extend to greatest depths below major mountain zones:

Fig 132 Relationships between surface features and crustal structure

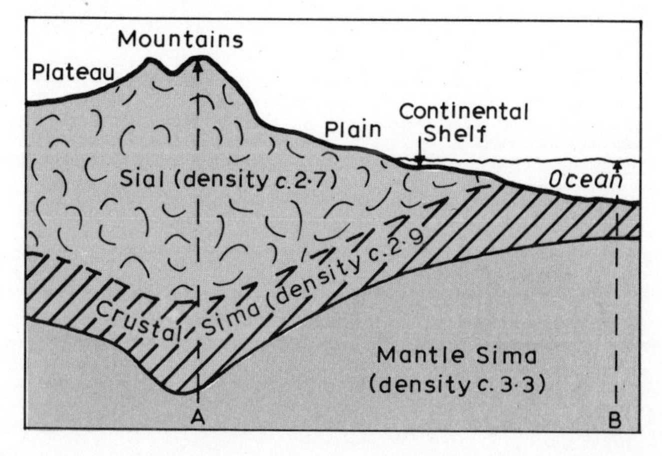

This means that the weight of material overlaying A (*Fig 132*) is approximately the same as that over point B, the greater volume above A being compensated for by the fact that more of the material is of the lower density.

Holmes calculates the *average* temperature at the base of the crust as 430°C—well below its melting point—but makes it clear that there are many areas where much higher temperatures occur. Although the mantle rocks are certainly not fluid (even if above their normal melting point, the effects of extremely high pressure would have to be allowed for) they appear to be able to 'flow' very slowly. Possibly these rocks are in a plastic state—a comparison has been drawn with hard toffee which breaks upon a sharp blow but which will bend under slow, continuous pressure. From this arises the concept of **isostasy**—the idea of continents 'floating' on the mantle rocks. If a set of different size wooden blocks are floated on water they will behave as in *Fig 133*. Block A is made of two parts and if the smaller part is removed and placed on B, the two blocks will float higher and lower respectively to adjust to the new conditions. This may be taken as a very rough guide to the behaviour of continental blocks following the erosion of mountain areas and the deposition of the eroded material as an alluvial plain. Similarly the addition of an extra block (*Fig 133c*) represents the effect of the accumulation of a great ice mass on part of a continent. When the additional weight is removed the land will rise to its original position—isostatic recovery. Of course this analogy is false in many respects: land masses are warped up and down rather than moving as separate blocks; and the whole process is extremely slow so that isostatic recovery after the last Ice Age is still going on.

In addition to isostatic adjustments, the rocks of the earth's crust are subject to major mountain building (**orogenetic**) forces. Possible mechanisms for such upheavals include convection current flows within the mantle. Such sub-crustal movements may have led to the drift of continental masses across the face of the Earth, to the thrusting up of new mountain chains and the down warping of great trenches (**geosynclines**) in which sediments accumulate to be later forced up in their turn into mountains. Such changes, always immensely slow, seem to have been concentrated at certain periods of geological time. These are indicated on the geological time scale (*Fig 1*) as the four major orogeneses which have effected the British Isles.

Rocks undergoing orogenesis are subjected to immense forces under which they may fracture (**fault**) or buckle (**fold**). Faults are often illustrated by means of fairly simple block diagrams (*Fig 134*) to illustrate the possible types of movement, but in nature faults are rarely as simple as this. The fault plane is not often a cleanly broken surface—frequently a fault zone of shattered rock can be observed on either side of the fault plane. It is the existence of such fault zones that has been referred to in the study above of Loch Linnhe. Faults may be of only local significance or they may be major features of the earth's structure, as for instance the Great Glen fault, or the San Andreas fault stretching 1300 km parallel to the Californian Coast, both predominantly transcurrent

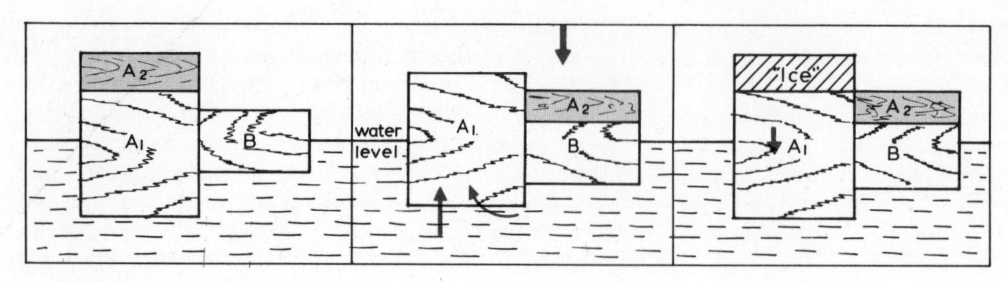

Fig 133 Diagram to illustrate principle of flotation as aplied to isostasy

or **wrench faults**. Land between parallel faults may relatively subside producing a **rift valley**, or may be relatively uplifted to form a **horst**. An example of the type of structure produced by block faulting is seen in *Fig 153*. An example of a fault line scarp is considered in *Study E4*, and aspects of fold structures are dealt with in *E6*.

FURTHER WORK

Practical

(1) Obtain photographs of fiords on the west coast of Scotland, Norway, New Zealand or British Columbia. From the photographs draw field sketches and annotate them to show

(a) evidence of glacial influence upon the landscape;
(b) physical features which have a significance for human geography, e.g. deep water near shore, deltas or raised beaches providing flat land for agriculture and settlement, plateaux or high flattish land for upland pasture, etc.

(2) Write a short comparison between the physical features of fiord and ria coasts as exemplified by Loch Linnhe and the Fowey Estuary (*Study B8*). If the O.S. sheets are available it is also valuable to compare the human geography of these coastlines.

Reading

Anderson: *Splendour of Earth*, pp. 339–342
Carter: *Land Forms and Life*, Section 29, pp. 226–231
Dury: *The Face of the Earth*, Ch. 5, pp. 59–60; Ch. 11, pp. 125–136
Gresswell: *Glaciers and Glaciation*, Ch. 3, pp. 41–44
——: *Geology for Geographers*, Ch. 1, pp. 9–13; Ch. 5, pp. 66–73
——: *Physical Geography*, Ch. 23, pp. 309–313
Hardy & Monkhouse: *Physical Landscape in Pictures*, pp. 80–81
Holmes: *Principles of Physical Geology*, Ch. XX, pp. 656–657; *Faults, Ch. IX, pp. 218–233; *Isostasy, Ch. II, pp. 27–30 and Ch. III, pp. 58–63
Horrocks: *Physical Geography and Climatology*, Ch. 9, pp. 147–149; Isostasy, Ch. 1, pp. 18–20
Monkhouse: *Principles of Physical Geography*, Ch. 10, pp. 255–257; Faults, Ch. 2, pp. 28–30; Isostasy, Ch. 2, pp. 17–19
——: *Landscape from the Air*, pp. 45 and 46
Scovel: *Atlas of Landforms*, pp. 138–139
*Sparks: *Geomorphology*, Ch. 12, pp. 277–280; Isostasy, Ch. 14, pp. 327–332
Wooldridge & Morgan: *An Outline of Geomorphology*, Ch. XXI, pp. 324–325; Faults, Ch. VII, pp. 86–96

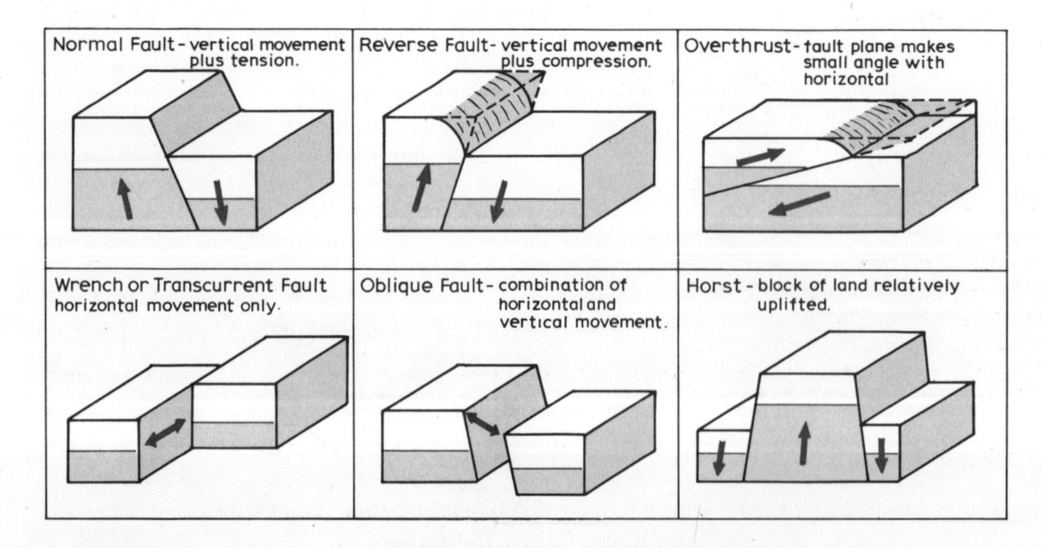

Fig 134 Types of faults

Section D: Denudation under Arid and Semi-arid Conditions

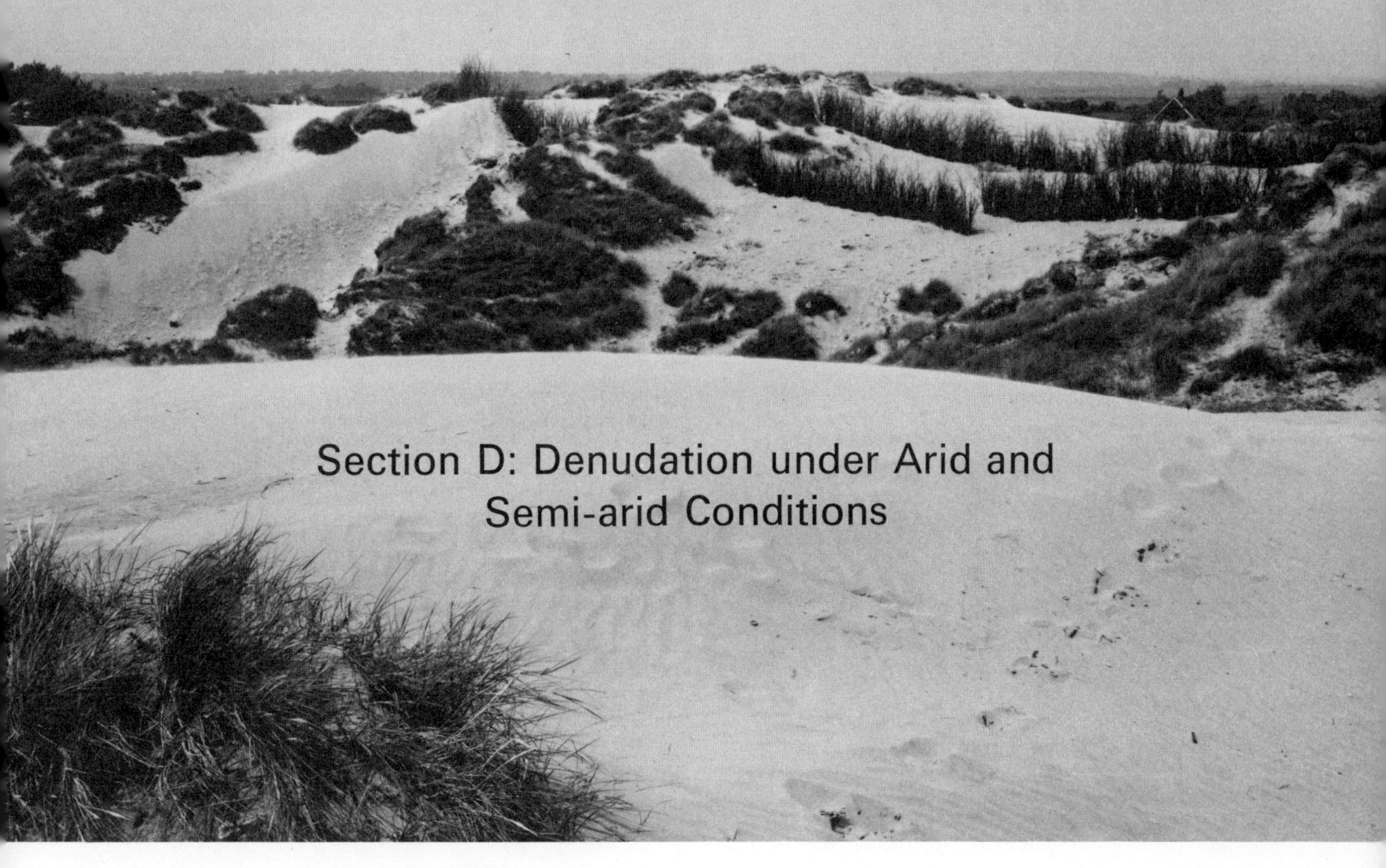

Fig 135 Wind blown sand dunes near Camber, Sussex

SAND AND WIND

Study D1

All the landforms considered in previous Studies have been and are being shaped by moving water either in its liquid form (soil water, rivers, the sea) or as a solid (ice). Where the climate is deficient in moisture a different balance exists in the processes of denudation, and landforms occur which are characteristic of arid areas. The transition from humid to arid climates is not sudden, so that there exist many intermediate stages. Increasing aridity is marked by a gradual deterioration in the vegetation cover from poor grassland and scrub with some bare patches of rock or soil, through a wide variety of semi-desert plant communities to the driest desert lacking all vegetation. Where bare soil and rock debris occur they are subjected to intense heat by day and rapid cooling at night due to the general absence of cloud cover, and further fragmentation of particles occurs (*p. 2*). More important, the absence of vegetation means that above the ground there is nothing to reduce wind velocity over the surface, and below the ground there are no root systems to bind the soil particles together; and in addition there is no soil moisture near the surface to give coherence to the particles. As a result the wind is able to play a more significant part in moulding desert landscapes than it does in wetter areas.

This does not mean that winds are necessarily stronger in deserts or that wind cannot, on occasions, modify landforms in humid areas. Farmers in Fenland and Lincolnshire found this to their cost when gales following dry spells in the Springs of 1968 and 1969, swept hundreds of tons of irreplaceable top-soil from their fields and piled it in drifts along hedgerows and roads. Many readers will have experienced a day on a sandy beach made miserable by blown sand stinging the body and forcing itself into every fold in clothing and every cranny of the picnic basket. The best studies of wind as a landscape forming agent in Britain can be made in areas of coastal **sand-dunes** (*Fig 135*). These have been piled up from loose sand deposited by the waves and blown inland to form lines of usually patternless hillocks which are then slowly colonized by vegetation communities until they are 'fixed' in a more or less stable form. Stability thus formed can, however, easily be destroyed if the plant cover is reduced by, for example, excessive trampling by tourists.

In truly desert areas there is usually no vegetation to fix the loose sand and the wind has free play. The processes by which wind moves sand and other particles are at least as complex as those involved in river action. Only extremely small particles, those with a diameter of

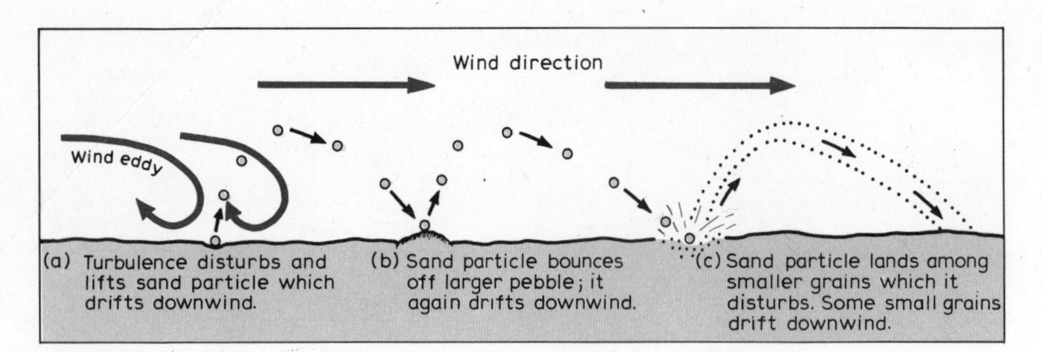

Fig 136 Downwind movement of sand grains

less than 0.2 mm which may be described as dust, can be carried freely for long distances and up to considerable heights by the wind—such material may be likened to the suspension load of rivers. Grains of larger size than this are moved in a series of jumps, each jump being initiated by turbulence concentrating the wind force sufficiently at one spot to lift some particles into the air and these then drift down-wind as they fall (*Fig 136*). When the particle lands it will either bounce upwards again after striking a larger particle or it will disturb and throw up smaller grains if it lands amongst them. In either case there will be a further upward and forward bound, thus resulting in a general downwind movement of material. The height and length of each forward bound will vary with particle size and wind speed, but it is rare indeed for particles undergoing this process, comparable to saltation in rivers, to be lifted more than 2 m above the surface. In addition to the bounding progression described above, falling grains may strike against larger particles on the surface disturbing them sufficiently to initiate a downwind surface creep.

In view of the foregoing facts it must be clear that any erosion by wind, or more accurately by the sand grains carried by the wind, must be limited to within at most 2 m of the surface. Furthermore, if a relatively small obstruction such as a boulder stands above the surface, the general air flow will be diverted around it thus reducing the impact of the sand grains. Certain steep rock faces are undercut, rock surfaces may be polished or irregularities in their structure etched out, but it seems likely that few landforms can be attributed solely to the erosive effects of wind-borne sand grains.

Of greater significance is the function of the wind in removing the finely divided products of weathering. The lower limit of this blowing away—known as **deflation**—is the water table since permanently moist material will have too great a coherence to be blown away, and it seems likely that some hollows or depressions in desert areas can be attributed in part to wind action. Deflation also tends to sort out desert material according to size, producing areas of stone or pebble covered surface (**hamada** in the Sahara) and other areas of 'sand seas' (the **erg** of the Sahara), and carrying right out from the desert the finest dust which becomes washed from the air by rains on the desert margins to accumulate as **loess**. The loess of North China is derived from the Central Asian deserts; loess or 'limon' soils in Europe, e.g. in the Paris Basin and along the S edge of the North European Plain probably owe their origin to finely ground material blown outwards from the edges of the retreating ice sheet of the last Ice Age. Occasionally fine Saharan dust even reaches the British Isles carried by winds at high altitude

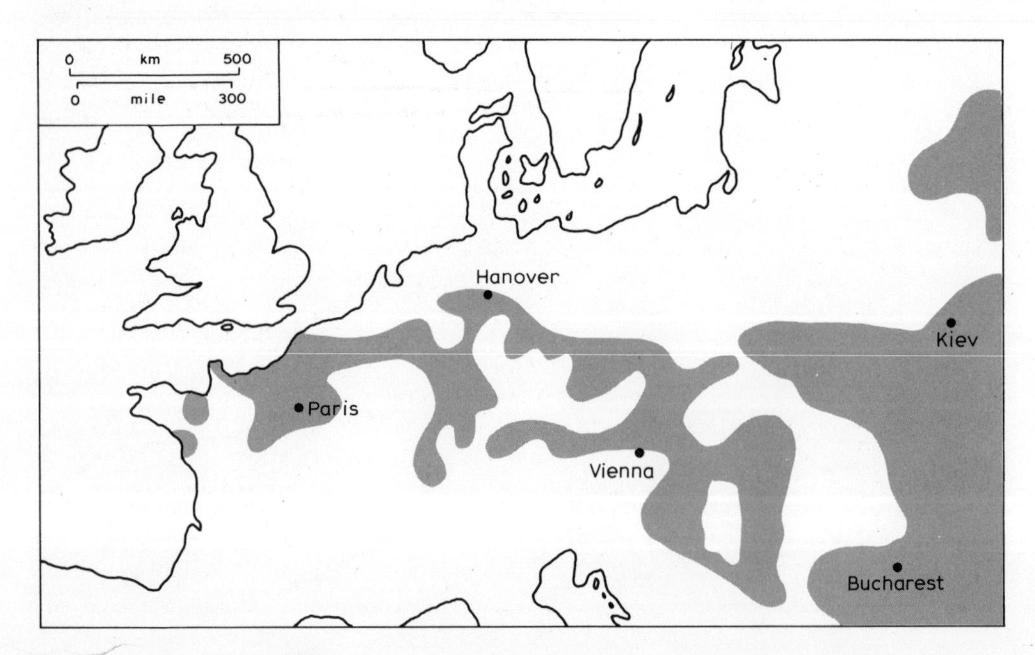

Fig 137 Approximate extent of Loess (Limon) in Europe. N.B. There is evidence that some soils in S. England have wind-blown constituents related to loess

Fig 138 Approximate extent of loess in North China

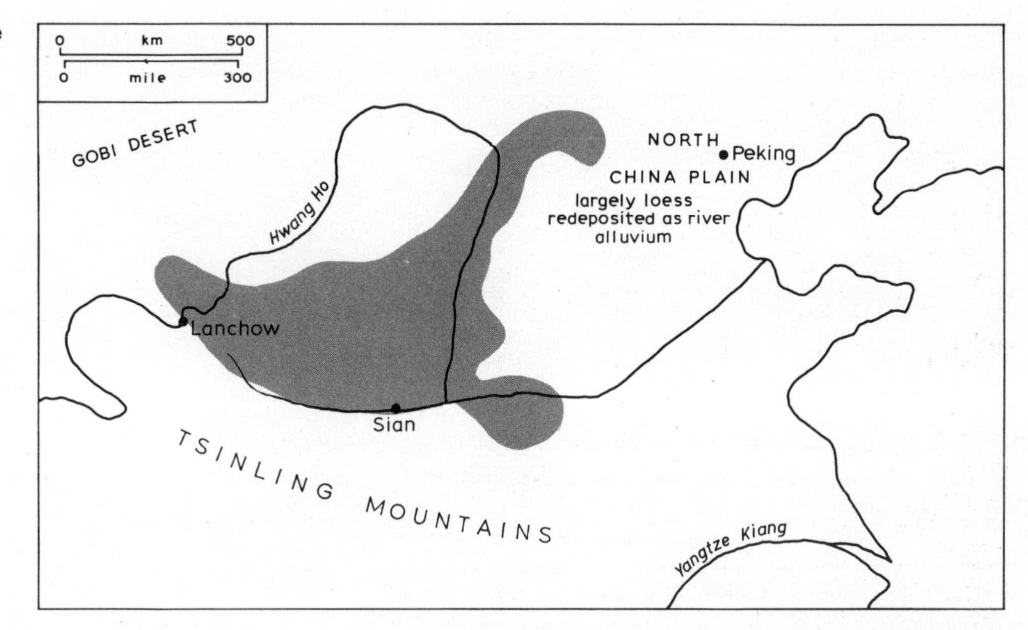

and washed downward by rain. Such an event was widely reported in the press on a day in July 1968, when the dust formed a coating on cars and stained washing hung out to dry.

Loess represents one feature attributable to deposition by the wind, and it is as an agent of deposition that the wind produces some of its most obvious effects upon the landscape. Accumulations of sand can occur in association with marked surface irregularities such as a large boulder or a gap in an escarpment. In the lee of such features wind flow is reduced either by eddying in the case of an obstruction (*Fig 139a*) or the spreading out of a wind current which was funnelled along a gap (*Fig 139b*). As **sand shadows** or **sand drifts** formed in this way increase in size by further accumulation some of the excess sand is caught by the wind and blown away, so that the feature remains more or less constant in size though continually changing in composition. Of course such features are subject to modification in size and shape, and perhaps also to complete removal, if the wind speed and/or direction changes markedly.

A very characteristic depositional feature of the sand covered deserts (which incidentally comprise less than one-quarter of the total desert area of the world) is shown in *Fig 140*. Light from the low sun emphasizes the moulding of these dunes, with a relatively gentle slope towards the left—the direction from which the dominant winds blow—and a much steeper leeward slope. The dunes also exhibit a tendency towards a crescentic plan. When completely developed a dune of this form is known as a **barchan** (or barkhan). The mechanics of dune formation are complex. It has been shown for example that barchans can only begin to form when sand of suitable size (0.5 mm to 0.125 mm in diameter) collects in piles larger than a critical height of about 0.3 m. The accumulation of such a pile may be the result of long continued action by a gentle wind, while the actual moulding of the barchan can be the work of stronger winds from a different direction. But whether this is the case or not, once the critical height is reached dune forming winds begin to propel sand grains up the windward slope. As this sand accumulates on the top of the pile, the lee side steepens

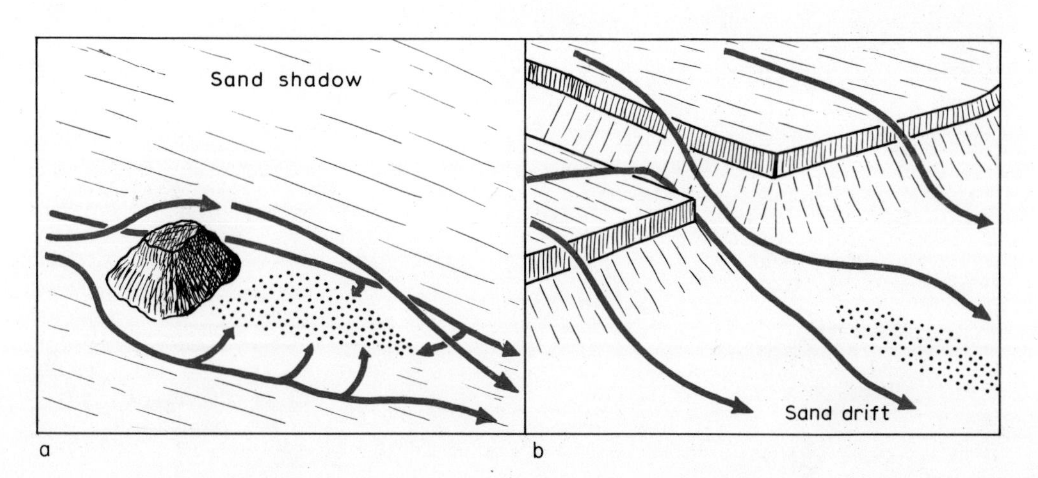

Fig 139 Wind blowing over and around an obstacle, depositing a 'sand shadow' on the lee side (a)

Sand drift deposited as wind spreads out after being concentrated in a valley (b)

Fig 140 Sand dunes in desert valley, Utah

and becomes unstable until sand slips down leaving a leeward slope of approximately 34° which is the angle of rest for dry sand grains of dune forming size. The edges of the pile, being smaller, are moved forward rather more quickly thus producing a lee slope which is concave in plan—hence the crescent shape of barchans. It will be clear from the preceding description that, with sand constantly moving up the windward slope and falling down the lee slope, a barchan moves slowly forward (*Fig 141*). The rate of this advance is in part dependent on size—a barchan 3 m in height advances about 18 m per year; one 18 m in height moves forward about 10 m per year, and dunes 30 m high (the maximum recorded) are nearly stationary. The maximum recorded width of barchans is just over 350 m. Barchans of true symmetrical form will only occur when winds of dune forming strength blow solely from one direction: more variable winds will distort the pattern (*Fig 142*).

The other characteristic dune type is the **seif**, or longitudinal dune. These are ridges of sand parallel to the wind direction up to 90 m in height and often extending for more than 80 km. The distance separating crests of such lines of dunes ranges from 18 to 450 m.

Fig 141 Forward movement of barchan (in cross section)

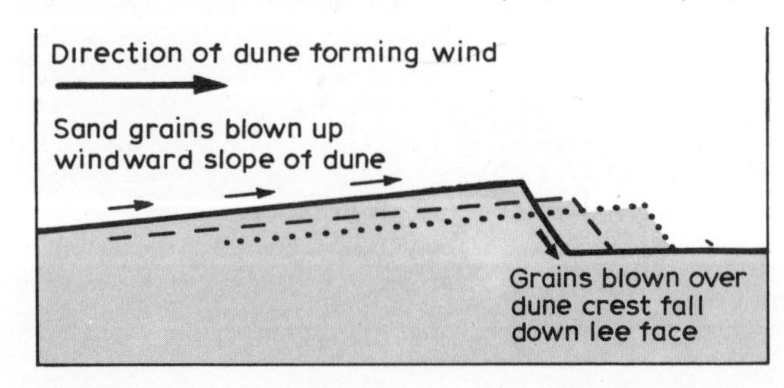

FURTHER WORK

Practical

(1) Make small piles of very dry sand of different grain sizes and, by removing a little sand from the base of each

Fig 142 Distortion of barchans under condition of changing wind direction

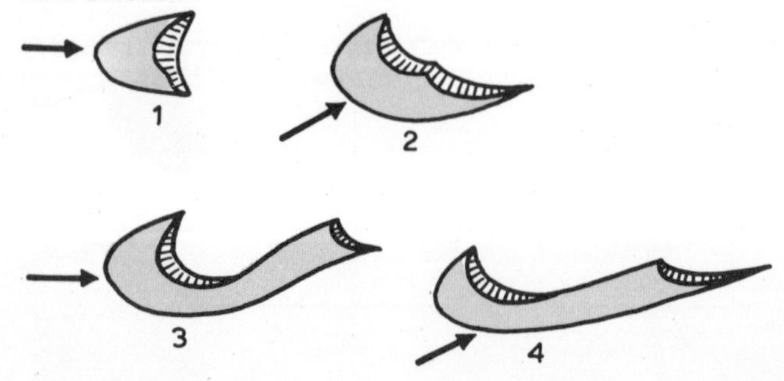

pile, determine the **angle of rest** for each type of sand. Repeat with moist sand and note the differences.

(2) Possible observations on a dry sandy beach on a windy day:

(a) Can you determine the maximum height to which sand grains are lifted by the wind?—a pole bound at 0.5 m intervals with Sellotape (sticky side outwards) will trap airborne sand.

(b) Try to observe closely how sand grains of different sizes are moved over the surface of the beach.

(c) Can you observe any tendency for the sand to accumulate with a gentle windward slope and a steeper slope on the lee side?

(3) If an area of coastal sand dunes is accessible, make careful observations by field sketches or photographs of any evidence of sand accumulation (partially buried vegetation, notice boards, etc.) or of recent erosion following any destruction of the plant cover. Do old maps or any other sources of information suggest an increase or decrease in the extent of the dunes?

Reading:

Dury: *The Face of the Earth*, Ch. 16, pp. 192–196

Gresswell: *Beaches and Coastlines*, Ch. VI, pp. 87–94

——: *Physical Geography*, Ch. 29, pp. 389–399

Hardy & Monkhouse: *Physical Landscape in Pictures*, pp. 64–67

Holmes: *Principles of Physical Geology*, Ch. XXII, pp. 748–768

Horrocks: *Physical Geography and Climatology*, Ch. 7, pp. 114–118, 122–124

Monkhouse: *Principles of Physical Geography*, Ch. VIII, pp. 160–170

——: *Landscape from the Air*, p. 36

Scovel: *Atlas of Landforms*, pp. 146–155

Simons: *Deserts*, Oxford Univ. Press, 1967

*Sparks: *Geomorphology*, Ch. 11, pp. 245–253

Strahler: *Physical Geography*, Ch. 28, pp. 434–447

Wooldridge & Morgan: *An Outline of Geomorphology*, Ch. XX, pp. 271–279

WATER IN THE DESERT — Study D2

At the beginning of the previous Study certain differences between conditions in humid and arid regions were described. One major difference affecting landforms was not however considered—namely the fact that lack of precipitation combined with high rates of evaporation prevent the formation of permanent rivers in deserts. There are, of course, a few examples of rivers which rise outside a desert and have sufficient volume to flow perennially right across the desert region—the Nile and Colorado are obvious examples. But in deserts and semi-deserts the possible evaporation is greater than the precipitation, even if only for part of the year, and thus few rivers with sources in arid or semi-arid regions can flow perennially, and those which do flow lose so much water by evaporation that they dry up before reaching the sea. It is characteristic of such regions that they are areas of **interior drainage**. In some places the drainage centres on lakes or inland seas in which the inflow of water is roughly balanced by the loss by evaporation, e.g. Dead Sea, Caspian Sea, Great Salt Lake.

(*Problem to solve:* Why are such lakes saline?)

Alternatively the centre of a basin of interior drainage may normally be a dry or marshy salt-encrusted expanse which only becomes a true lake of standing water when its feeder channels bring down flood water after heavy rain. Such a lake is Lake Eyre in Central Australia; all temporary lakes of this kind are known as **playas**.

Permanent streams are thus the exception in arid and semi-arid regions and ground water is very limited indeed. Hence the processes of river action and soil creep, fundamentals of landscape formation in humid lands, are not to be expected to occur in the same form in the world's deserts. But the action of running water can by no means be discounted in interpreting arid landscapes for three reasons:

(a) Towards the fringes of the desert proper and in semi-desert regions, the climate is frequently such that evaporation exceeds rainfall for only part of the year. In other months streams can flow and considerably modify the landscape.

(b) Dry regions characteristically have very unreliable rainfall and even the most arid areas may, very occasionally, experience torrential rain—e.g. Trujillo, Peru, received 35 mm between 1918 and 1924 and then 380 mm fell in the month of March 1925. This may be an exceptional case, but other desert areas show records of extremely heavy rain concentrated into a short period. As will be shown below, the effects of such storms upon landscape formation is likely to be considerable.

(c) There is ample evidence, particularly in the case of the Sahara, that as recently as early historic times certain areas which are now true deserts experienced a semi-arid climate, and that present-day semi-deserts were regions with sufficient rainfall to be classified as humid. This change has taken place in such a short time, geologically speaking, that it is inevitable that the present landscape will contain relics on a large scale of the scenery produced under conditions of higher rainfall.

It is characteristic of arid climates that such rain as does fall occurs in the form of convectional storms of high intensity. The run-off is thus very rapid and flood water pours in great volumes along normally dry watercourses.

Fig 143 Rain clouds and desert flooding

During the probably long period since a previous flood the land has been mantled with the finely broken products of desert weathering and wind deposition, so that the torrent is provided with enormous quantities of small particles which can be easily transported, and also with larger stones which can be rolled along the stream bed. While such a torrent is confined within a pre-existing valley its erosive power is likely to be considerable, and such storm torrents are probably the major erosive agents in the deserts at the present time. As the volume of water declines with the cessation of rainfall, and with rapid loss by evaporation and percolation, a sheet of debris is left covering the floor of the watercourse to be later redistributed by the wind. Valleys formed or enlarged by temporary flood water torrents in deserts are termed **wadis** and are characteristically flat floored with steep rocky sides. This form may be explained by the fact that the torrent spreads across the whole width of the valley and by the fact that the water is not available for long enough to allow soil-creep and weathering to reduce the slope of the valley sides. It may also be that, in periods between torrents, some wind erosion undercuts the valley sides thus accentuating their steepness. It is useful at this point to reconsider the sand-pile experiments suggested in *Study A3*. Because of the single water-source used, the unconsolidated nature of the sand, and the absence of weathering, the type of processes operating were in fact very similar to those involved in wadi formation.

By no means all wadis observed in deserts today can be attributed solely to contemporary conditions. The close spacing of such valleys often indicates that they must have originated when the rainfall was higher, and some contain caves from which springs would have emerged in more humid conditions.

When a flood torrent flowing down a wadi reaches the edge of the highland area, its waters are not confined by the high ground on either side of the valley. As the water, heavily loaded with rock debris, spreads out and its energy becomes less concentrated, its load carrying capacity is reduced. Coarser material is deposited first leading to the division and sub-division of the stream, i.e. it becomes braided. With progressive division, spreading out and reduction of stream size, and loss of water by downward percolation and evaporation, smaller particles are deposited at lower levels and an **alluvial fan** is produced. Alluvial fans at the foot of mountain ranges are by no means confined to desert landscapes. They frequently occur in humid lands especially where seasonal floods are characteristic—see *Fig 45* (*Study A11*) where a fine set of such fans can be seen at the foot of the mountains in the background of the photograph (*also Fig 146*).

In arid areas the flood torrents will sometimes carry sufficient water for the stream to extend across the alluvial fan and carry water to the centre of a basin of interior drainage, where a temporary playa lake is formed. Finely divided material will also be carried in to the centre of the basin. The evaporation of the waters of a playa lake leaves behind salt particles which can cement together the sand grains and, over long periods, the repetition of this process produces very level, firm surfaces such as the **salt flats** of Utah which have provided ready-made tracks of sufficient strength for attempts upon the world land speed record.

A feature of desert and semi-desert landscape which has aroused much discussion among geomorphologists is the existence at the foot of rocky outcrops of a gently sloping platform of bare rock known as a **pediment**. Formed of the same rock as the highland, the pediment is distinguished from it by a sharp break in slope. Further from the highland the pediment becomes lost beneath an increasing thickness of deposited material known as the **bahada** or **bajada**, which in turn merges into the finely divided particles covering the centre of the playa basin. Linked with the problem of explaining pediment formation with its sharp break of rock slope, is the occurrence of isolated, steep sided island mountains—**inselbergs**—standing out above a general plateau surface. The detailed forms of inselbergs, pediments and bahadas will vary according to climatic conditions and the characteristics of the bedrock. Thus granitic rock in semi-arid conditions

Fig 144 Granitic inselbergs, Moẓambique (after Holmes)

may give rise to a rounded inselberg form with an extensive pediment of almost bare rock (*Fig 144*); whereas horizontal sediments in more arid conditions are likely to exhibit flat-topped blocks known as **mesas** with a pediment largely masked by rock debris (*Fig 145*).

Fig 145 Basalt capped mesa, Texas

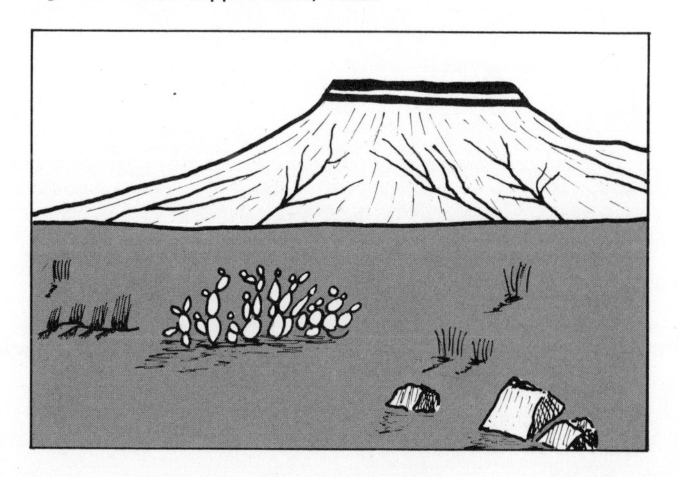

Fig 146 Alluvial fan at edge of Death Valley, California

Fig 146 is a photograph of part of the edge of Death Valley, California. Careful study shows that there are three types of surface in the area illustrated. The almost straight slope cut into solid rock meets the virtually horizontal surface of the valley floor at a sharp break in slope in the middle distance. What is the angle of the rock slope? Not only does the angle of slope change sharply, but so too does the nature of the surface; from solid rock to finely divided rock waste. So the break in slope marks a change in the dominant process, from erosion wearing away the rock slope to deposition where rock waste is accumulating in the valley floor. Any pediment cut into solid rock is here covered by an accumulation of rock debris.

Closer to the camera an additional feature can be seen between the rock slope and the flat valley floor. This is an alluvial fan with a slope angle less steep than the rock slope above but not as flat as the valley floor. Close study of the alluvial fan shows minor gullies crossing it, mostly radiating outwards from the highest point of the fan. The rock slope itself, while generally straight, is also diversified by gullies two of which can be seen to emerge near the apex of the fan. Despite the normal aridity of this area, it is clear from this evidence that erosion of the rock slope is by running water after the occasional torrential rains. The waste products of this erosion are washed down the gullies and build up at the foot of the slope as an alluvial fan where the reduced speed of stream flow and the increasing volume of load leads to deposition. Following exceptionally heavy rain some of the

Fig 147 Death Valley, after a storm

finest rock debris will be carried out on to the valley floor, and it is here that wind action assumes significant proportions in redistributing the finer particles and perhaps forming dunes.

Evidence that such exceptional rainfalls occur is provided in *Fig 147*, also of Death Valley, where a temporary shallow lake may be seen to cover part of the valley floor following a storm whose departing clouds lie over the mountains in the background. In this photograph the contrast between rock upland and valley floor is again noticeable, and three alluvial fans can be identified.

Theories concerning the origins of pediments and inselbergs rest on very varied and often apparently contradictory evidence, and there is no general agreement as to the dominant process in their formation (see 'Reading', especially Dury, Holmes and Sparks). In general however it is agreed that inselbergs represent the final remnants of previously more extensive upland areas, and their study has led some geomorphologists to believe that denudation of an upland commonly proceeds by the parallel retreat of slopes rather than by the overall lowering of highland and that this is true not only of arid and semi-arid regions but of humid lands also.

The upper set of block diagrams and generalised cross section on the opposite page (*Fig 148*) represents the sequence of landscape evolution proposed by W. M. Davis which has been accepted for more than half a century as the normal pattern for landscapes affected by river action (*cf. Fig 44, p. 30*). Under this scheme an initial period of vertical erosion by rivers (Stage A) is continued through Stage B at which point the last remnant of the original surface remains. From this point onwards the divides are lowered more rapidly than the valley floor and the angle of slope thus becomes progressively less steep. A further point is that throughout the later stages of the cycle (from Stage B onwards) slopes retain a convex-concave form. This means that rounded forms predominate after the initial stage, and that sharp changes of slope angle (breaks of slope) are exceptional features requiring special explanation.

The alternative sequence, illustrated in *Fig 149*, is based in part upon the observation of landscapes formed largely by the action of running water in semi-arid regions such as were described on the preceding page and which were seen to exhibit a sharp change in slope angle between the hillside cut in rock and a plain of alluvial deposition. Support for this sequence is also claimed from the results of the increasing volume of detailed slope surveys in many parts of the world. Four elements in the slope from interfluve to river are recognised:

the **waxing slope** at the top, curving over to join the near vertical **free face** cut into solid rock. Below this is the less steep **constant slope** which is a straight slope which may be partly in solid rock and partly formed of rock debris and soil (**the regalith**). Finally there is the **waning slope** which in large measure is the alluvial filled valley floor.

Apart from this recognition of various slope elements, the sequence differs from that of W. M. Davis in visualising that the main slope elements remain constant in angle while being worn back. That is to say, in Stage C (*Fig 149*) the waning slope of the valley floor is becoming steadily more extensive and the waxing slope or interfluve is being reduced in extent, but that the angles of slopes (free face and constant slopes) between these two limits remain constant.

The protagonists of this theory suggest that the undulating, rounded landscapes of, for example, North-western Europe and North-eastern United States are exceptional in that they owe their origins to the intensely cold climates of glacial times and are not the normal result of stream erosion as has previously been assumed. Prolonged research will be necessary before the final answer to this major controversy can be determined.

FURTHER WORK

Practical

(1) Repeat the sand pile experiments of *Study A3* in such a way as to produce an alluvial fan. Make careful sketches of the stages in its formation.

(2) Study various textbooks of Climatology to discover what statistical limits have been proposed for the classifications of Desert and Semi-desert.

(3) Many students will wish to follow their reading on the theories of parallel slope retreat by discussing the implications of these ideas upon our understanding of landscape formation—to borrow Dury's chapter heading: 'Wearing down or wearing back?'

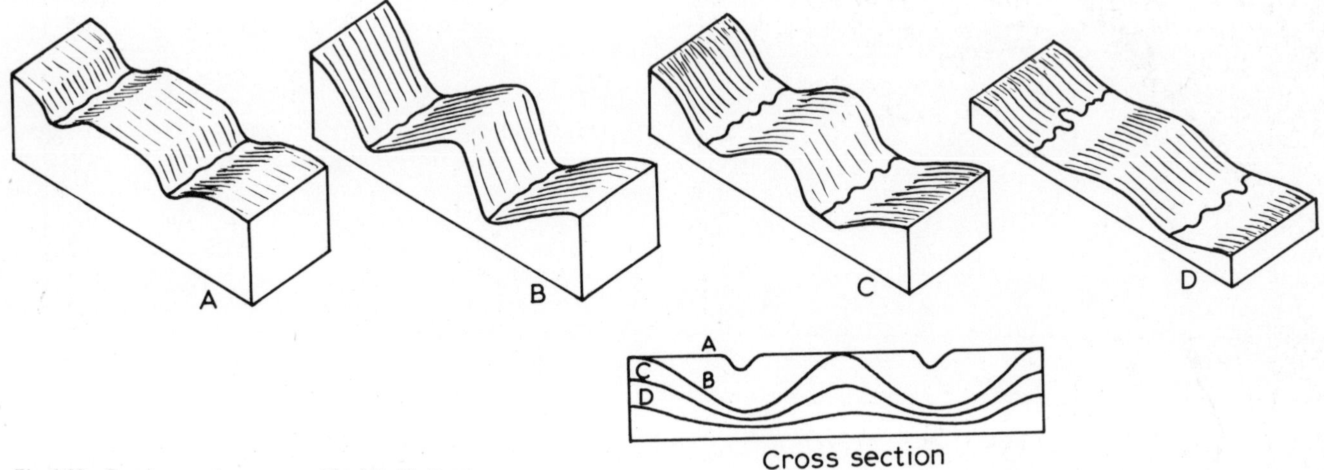

Fig 148 Erosion cycle proposed by W. M. Davis

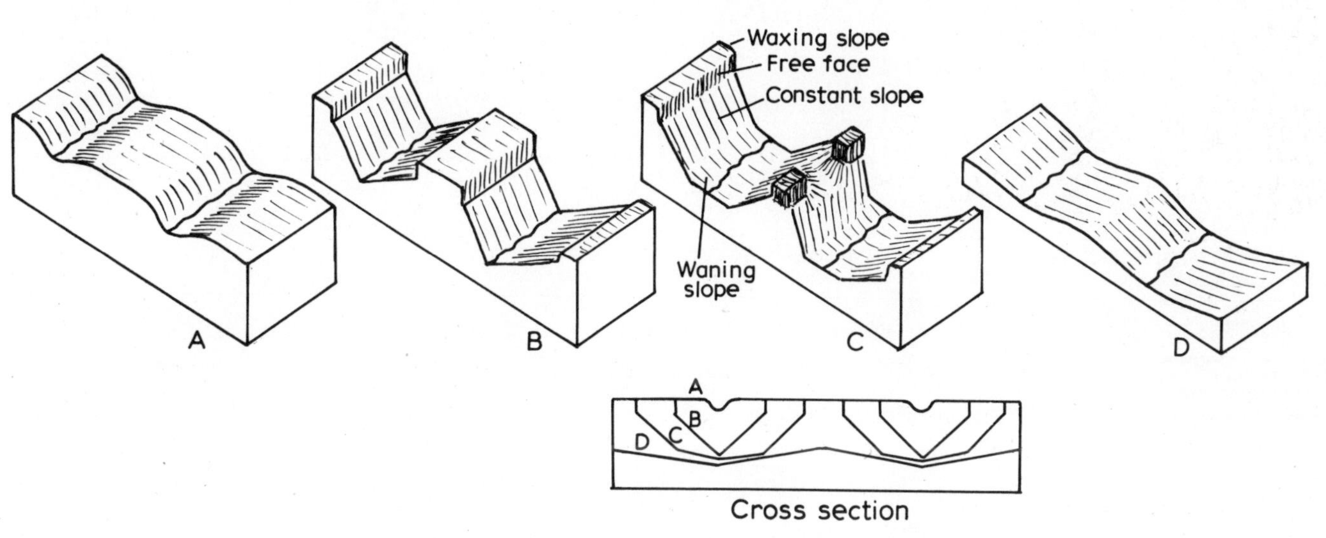

Fig 149 Erosion cycle based on parallel retreat of slopes

Reading

Dury: *The Face of the Earth*, Ch. 16, pp. 189–192
Gresswell: *Physical Geography*, Ch. 14, pp. 124–130
Hardy & Monkhouse: *Physical Landscape in Pictures*, pp. 65–67
Holmes: *Principles of Physical Geology*, Ch. XXII, pp. 768–776; *Ch. XVII, pp. 474–479; *Ch. XIX, pp. 609–615
Horrocks: *Physical Geography and Climatology*, Ch. 7, pp. 118–124
Monkhouse: *Principles of Physical Geography*, Ch. VIII, pp. 170–172
——: *Landscape from the Air*, pp. 33–35
Scovel: *Atlas of Landforms*, pp. 8–11; 19; 46–47
*Sparks: *Geomorphology*, Ch. 4, pp. 55–75; Ch. 11, pp. 253–262
Strahler: *Physical Geography*, Ch. 24, pp. 356–359; Ch. 25, pp. 380–384

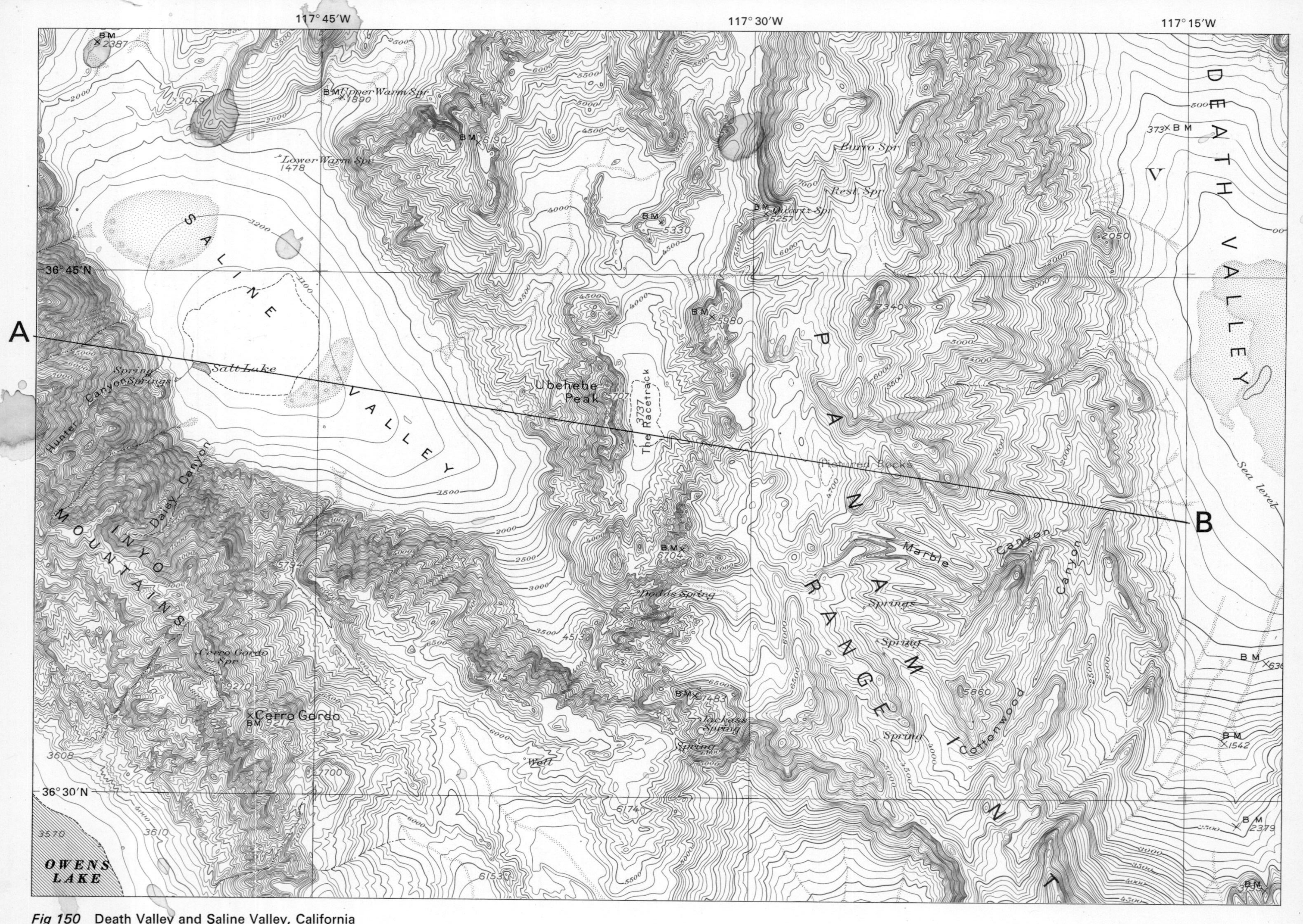

Fig 150 Death Valley and Saline Valley, California

Scale 1 : 250 000 (1 cm to 2.5 km *or* ¼ inch to 1 mile)

Region between the Sierra Nevada and Death Valley, California

Study D3

RANGES AND BASINS IN AN ARID REGION

U.S. Geological Survey, Ballarat Quadrangle
1 : 250 000 (1 cm to 2·5 km *or* approx. ¼ inch to 1 mile)

MAP EXERCISE:

(1) Locate this area on an atlas of the Western U.S.A.—the small water area in the SW corner is part of Owens Lake and the lowland in the E is part of Death Valley.

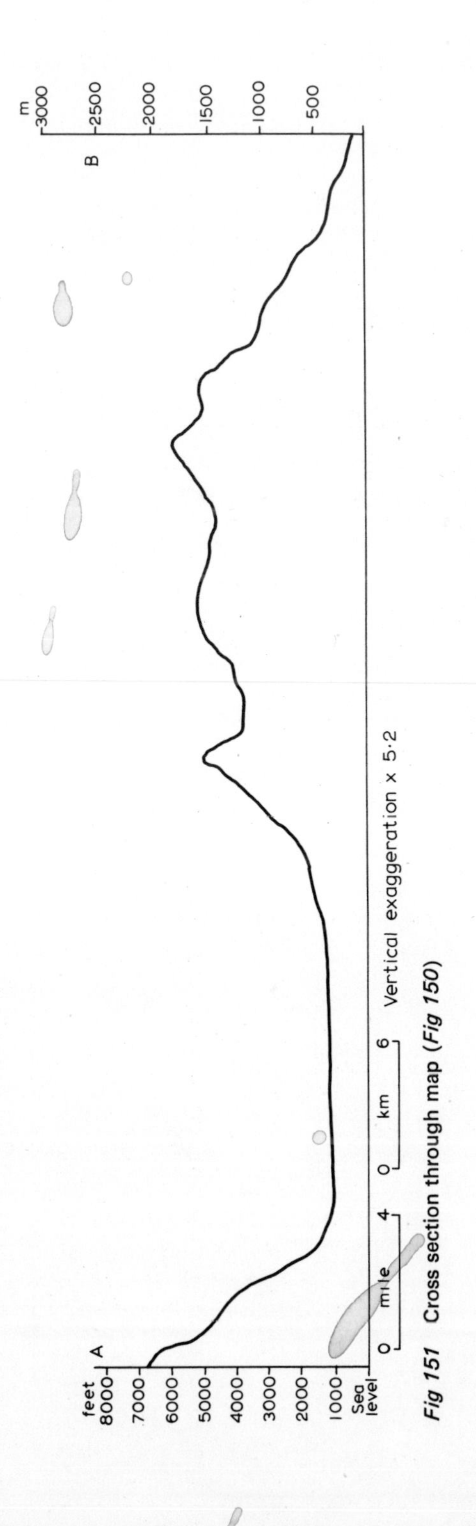

Fig 151 Cross section through map (*Fig 150*)

(2) What is the area covered by this map extract in km² and in square miles? Compare the size of the area shown with your home area as represented on the nearest equivalent O.S. map.

(3) *Fig 151* is a contour section from A to B—copy this diagram and annotate it to show: Inho Mts. (part of the Sierra Nevada); Saline Valley; Ubehebe Peak; The Racecourse; Pictured Rocks; Panamint Range; Death Valley. Indicate the positions of temporary lakes (enclosed by pecked black lines on the map).

(4) What map evidence do you find that this is an area in which evaporation is greater than precipitation?

If one considers carefully the case of the temporary lake at Pictured Rocks, it can be seen that the lake basin is totally enclosed by the contour for 4700 ft (1430 m). The small amount of precipitation on the surrounding hills will run down-hill into this basin, and after heavy rain the lowest land will become a lake. Since there is no surface valley leading downwards from the basin the water can only disappear in two ways:

(a) by underground seepage;
(b) by evaporation.

The general absence of streams on the map extract; the numbers of dry valleys without, or with only temporary, streams; the disappearance of even temporary streams into alluvial spreads on reaching lower ground; and the wording 'Salt Lake' in Saline Valley; all these facts suggest aridity and this can easily be confirmed by reference to climatic maps in an atlas. If precipitation was greater in this area than it is, one could visualize the Painted Rocks lake expanding in area and deepening until its extent was marked first by the 4700 ft contour (1430 m) and progressively enlarging up to the level of 4900 ft (1495 m). At a little above this level the water would begin to overflow from the NW corner of the basin towards the hollow of The Racecourse nearly 1000 ft below. The Racecourse 'lake' in turn would enlarge and overflow to the SW into Saline Valley.

The basins and ranges in this area owe their origin to **block faulting** (*Fig 153a*). The differential uplift and depression of the blocks produced a series of ridges, uplift blocks (horsts) and isolated depressions at various levels. The erratic but torrential rainstorms of the present generally arid climate, allied perhaps to past periods of somewhat higher rainfall, carved out gullies in the mountain sides during the short but powerful life of run-off streams. These gullies were in some cases enlarged into canyons and wadi-like features—the whole sculpturing process producing the fretted **badland** topography now characteristic of the steeper slopes. The temporary streams so formed flow towards the centre of each enclosed basin which thus becomes the centre of an area of interior drainage. As the torrents reach the gentler gradients of the basin floors they divide and subdivide, moving more and more slowly and depositing their considerable load in the form of alluvial fans. Several of these may be seen at the points where streams reach the S and W side of Saline Valley. As the water spreads out into smaller and shallower channels it is subject both to rapid evaporation and to a high degree of percolation into the coarse, unconsolidated debris of the alluvial fan, but after an exceptionally heavy storm some water may remain to spread out into a lake in the centre of the basin, bringing with it a quantity of more finely ground rock waste. Such a playa lake evaporates in this arid climate leaving an encrustation of salt (salt flats) and sometimes a small, semi-permanent salt lake. The finer debris brought into the basin is subject to wind action as soon as it dries out so that sand drifts and sand-dunes may occur. On occasions which, at least under present climatic conditions are very rare, lake level may rise sufficiently for water to overflow from one basin into another so that in theory at least debris will be progressively concentrated into the lowest basins.

It is important to realize that in this, and in other similar arid mountain regions, major relief forms are the consequence of structural processes modified by fluvial erosion, and that wind action is only responsible for relatively minor and superficial landforms. An area such as that depicted on the map extract and photograph can be seen on analysis to comprise three distinct types of scenery:

(i) intensely gullied mountain regions within which steep slope angles predominate;
(ii) regions of relatively gentle slope and with broadly smooth surfaces leading from the foot of the mountain areas outwards towards the centre of the depressions;
(iii) the perfectly flat floors of the playa lakes.

The division between the regions of types (i) and (ii) is very sharp being marked not only by an obvious change in the general angle of slope but also by the transition from exposed bedrock to an area usually thickly veneered with rock waste. From the foregoing study it will be clear that in such arid mountainous regions the main changes which will occur with the passing of time are the slow reduction of the mountain ranges and the accumulation of the debris from this erosion which will gradually enlarge the basin areas until much of the land consists of debris-covered plains through which the remnants of the mountains will emerge.

Fig 152 Death Valley, Looking west toward the Panamint Range

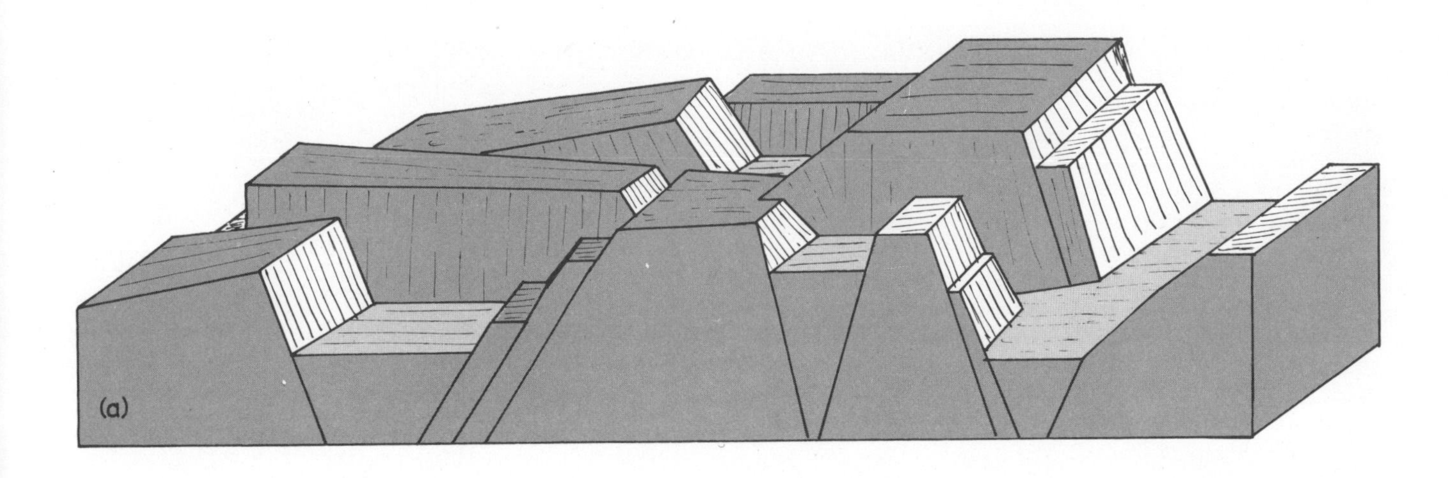

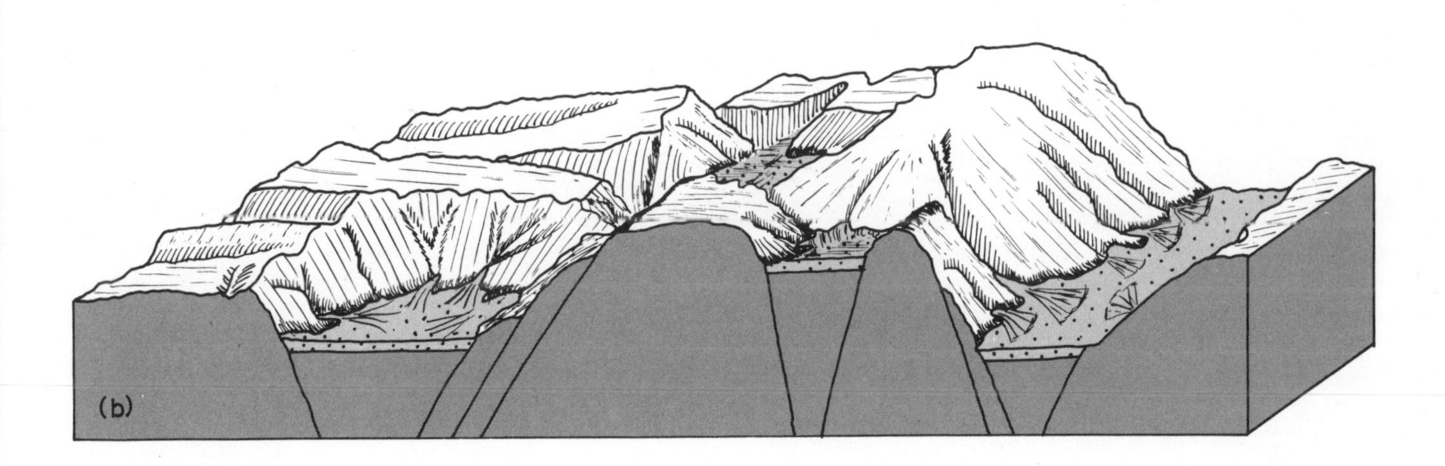

Fig 153 (a) Region of block faulting before erosion
(b) The same area affected by gulley erosion; note alluvial fans and deposition in basins

FURTHER STUDY

Practical

(1) Use an atlas to list the areas of the world which drain towards totally enclosed lakes or inland seas, e.g. Caspian Sea, Lake Eyre, Dead Sea. Draw sketch maps of two such basins and make a rough calculation of their areas.

(2) The map extract (*Fig 150*) shows a sea-level contour in Death Valley. From an atlas map find the altitude of the lowest point in this valley. What does the existence of land below sea-level in the middle of a continent imply concerning the structural processes which have operated in this area?

(3) Discuss the relative importance of wind and water in the formation of desert land forms.

(4) Make a tracing of *Fig 153b* and annotate to show the position of physical features mentioned in Studies E2 and E3.

Reading

Anderson: *Splendour of Earth*, pp. 371–376
Carter: *Land Forms and Life*, Section 12, pp. 90–95
Dury: *The Face of the Earth*, Ch. 16, pp. 193–199
Hardy & Monkhouse: *Physical Landscape in Pictures*, pp. 65–66
*Holmes: *Principles of Physical Geology*, Ch. XXII pp. 768–776
Horrocks: *Physical Geography and Climatology*, Ch. 7, pp. 118–122
Monkhouse: *Principles of Physical Geography*, Ch. 9, pp. 225–230
——: *Landscape from the Air*, p. 12
Scovel: *Atlas of Landforms*, pp. 44–45
*Sparks: *Geomorphology*, Ch. 11, pp. 256–262
Strahler: *Physical Geography*, Ch. 25, pp. 380–384
Wooldridge & Morgan: *An Outline of Geomorphology*, Ch. XX, pp. 281–283

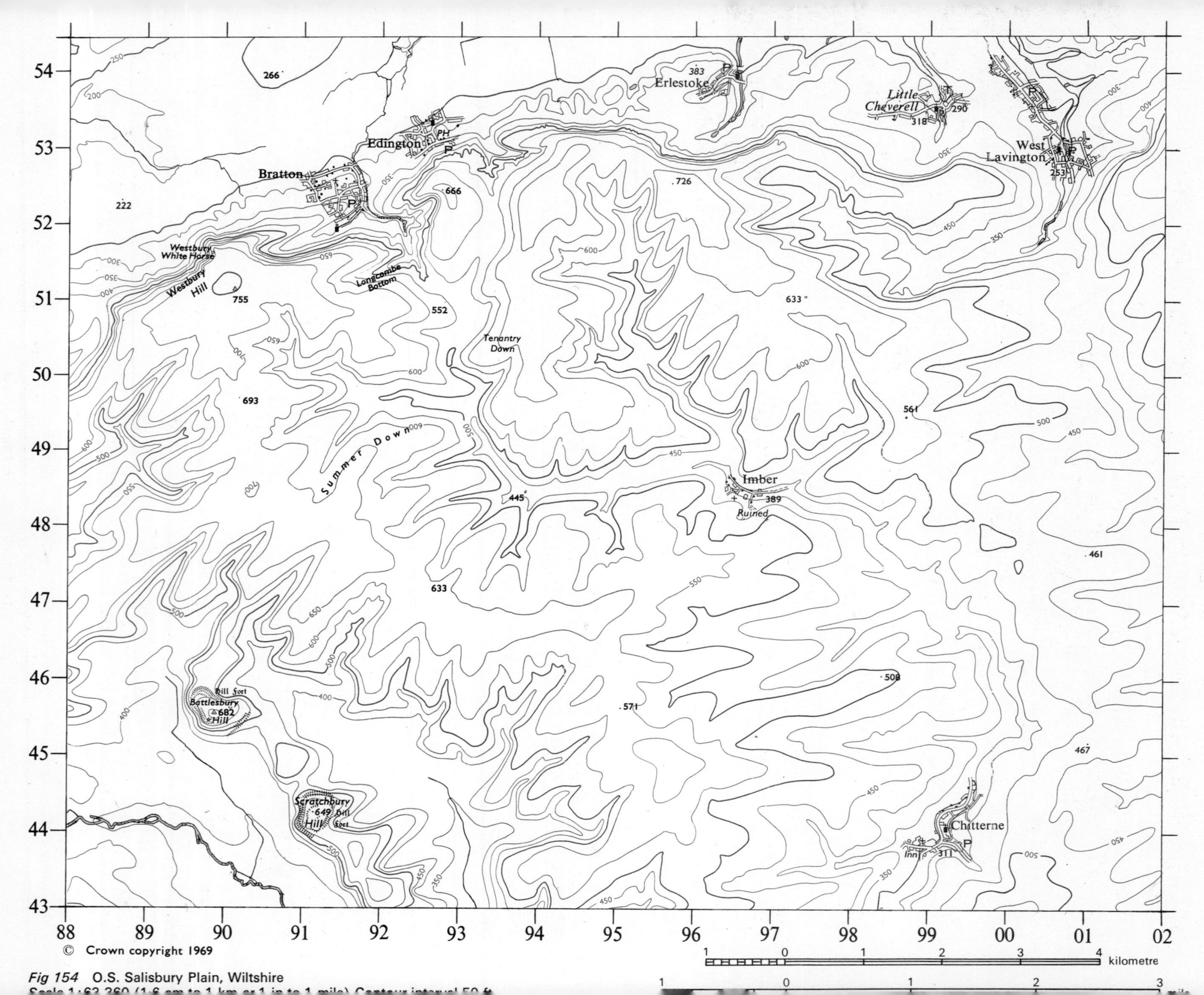

Fig 154 O.S. Salisbury Plain, Wiltshire
Scale 1:63 360 (1.6 cm to 1 km or 1 in to 1 mile) Contour interval 50 ft

Section E: Landscapes dominated by Rock Type and Structure

Fig 155 Chalk Downland, Woodhay, Berkshire

Salisbury Plain, Wiltshire

Study E1

A CHALK UPLAND

O.S. 1:63 360 (1·6 cm to 1 km *or* 1 in to 1 mile)
Sheet 167, Salisbury

In all the preceding Studies, emphasis has been laid on the **processes** which have contributed to the formation of landscapes. Many references have been made to the geological structure and to the physical characteristics of rocks as they affect the operation of the processes of erosion. Each rock type has qualities which modify the effects of erosion to produce a characteristic landscape. In the case of certain rocks and structures, this modification is so great as to warrant special consideration and in the sections which follow, some of these cases will be studied.

Fluvial erosion is dependent upon the movement of water over the surface of the land, with tiny rivulets joining to form brooks and eventually rivers. If, therefore, rainwater is rapidly absorbed by the underlying rock, fluvial erosion is reduced to a minimum and such rock is resistant to erosion irrespective of its physical 'hardness'. This explains why chalk, for example, although not a particularly hard or strong rock, forms areas of upland[1] rising above clay vales.

Chalk is a particularly pure, fine grained form of limestone made up of microscopic fossil shells embedded in a very fine debris of broken shelly material and calcium carbonate granules. Chemically it is almost entirely calcium carbonate and in Britain was deposited beneath clear seas of the Cretaceous period, and stratigraphically it is divided into Lower, Middle and Upper Chalk. The Lower Chalk contains more clay particles and is of a marly nature, and in the upper **horizons** (layers) of the Middle Chalk, and throughout the Upper Chalk, bands and isolated nodules of flint occur. Flint is almost pure silica, concentrated by solution into thin bands or knobbly pebbles; it is very hard and used by early man as tools because of its characteristic of fracturing to produce a sharp edge.

[1] The fact that chalk upland in Southern England is usually known as 'downland' must be accepted as one of the vagaries of the English language!

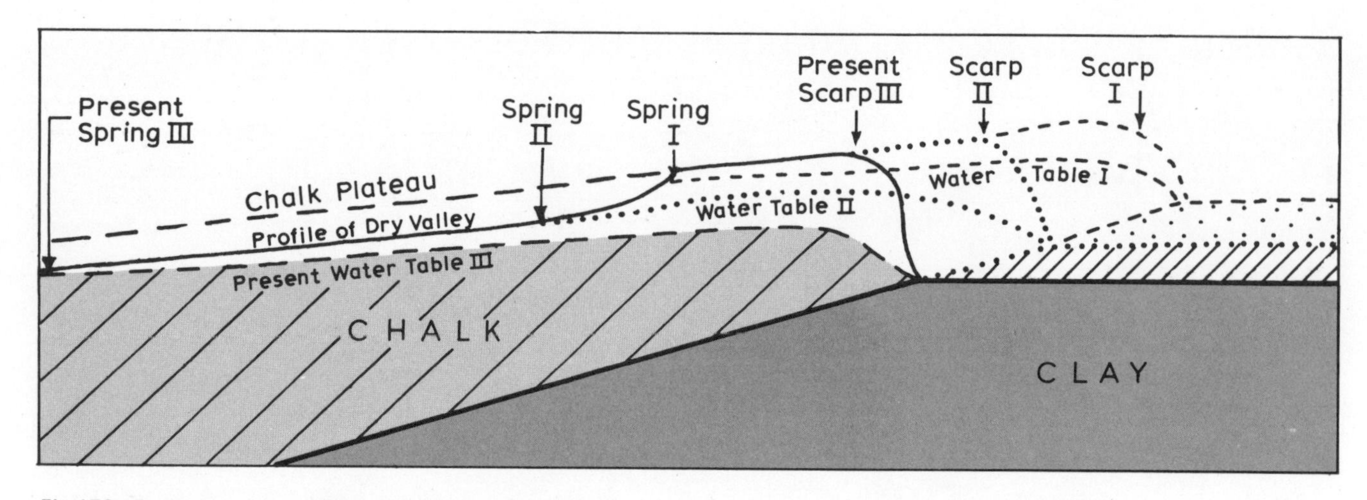

Fig 156 Formation of dry valleys by scarp recession (after C. C. Fagg)

Chalk is a **permeable** rock. That is to say it allows water to pass easily through it. It is also **porous**. These two terms are often confused. A rock is porous or non-porous according to the proportion of air spaces between the rock grains. A porous rock has a large proportion of such space, e.g. 20–30% in chalk, while a non-porous rock such as granite may contain less than 1% air spaces. But water moves only very slowly or not at all through these tiny spaces. Thus a rock may have 40% porosity, as do certain clays but because of the absence of continuous passageways it will not allow water to pass through it—it is therefore impervious or impermeable. Other rocks such as some granites or crystalline limestones have a very low porosity, but are broken by fissures and joints which permit the easy passage of water, and these rocks are thus permeable.

Chalk is both porous and permeable—it absorbs considerable quantities of water into its air spaces and allows water to pass easily through the rock because of the arrangement of particles and the existence of joints and fissures. The clearest indication in the landscape of the permeability of chalk is the absence of surface drainage. In the area covered by the map extract (*Fig 154*) and extending further E, one can easily delimit an area of 100 km² within which not a single stream occurs. Some dew ponds do occur but are, of course, man-made features. Their bed is lined with puddled chalk—that is chalk which has been mixed with water and compacted to render it impermeable, though modern dew ponds may have concrete lining.

Rain water percolates easily into the chalk until underlying impervious clay is reached. Water accumulates in the lower layers of chalk, filling the air spaces like a saturated sponge. The upper layer of saturation is the **water table**. Because water only moves slowly through the pore spaces and fissures, the water table tends to reflect the relief of the land (*see Study A4*), rising beneath the higher ground. When a well is sunk it will only yield water when it penetrates to the water table; even then it is often necessary to excavate horizontal headings to tap fissures in the chalk before enough water can be collected for an urban water supply or similar large scale demand. This emphasizes the fact that water does not percolate evenly throughout the chalk, but only moves freely along joints and fissures; it follows therefore that the water table cannot be regarded as a smooth surface, as usually shown diagrammatically. Such a concept can only be valid in the case of a completely uniform rock of high permeability, or as a rough generalization of conditions over a very large area of chalk countryside. A final point concerning the **hydrology** (water-bearing characteristics) of chalk is that springs occur only at well spaced intervals along an escarpment or dip slope (*see Studies A4* and *E2*). If water flowed evenly throughout the chalk mass one would expect to find an almost continuous line of seepages where the water table reached the surface. That this does not occur is further evidence that water movement is concentrated along fissures, etc. and it is where these reach the surface at a sufficiently low level that springs may occur.

The preceding account of some salient characteristics of chalk is a necessary preliminary to understanding certain features of the landscape of chalk downland such as that illustrated in the map extract. Study of the map indicates first the existence of a steep N facing slope rising from lowland of between 50 m and 75 m (200 ft and 250 ft contours) to a maximum height of 230 m (spot height 755 ft) at 901512. From the crest of this steep slope, summit heights decrease steadily in a SSE direction. If a 1 : 63 360 geological map is consulted it will be seen that the steep face marks the N limit of Upper and Middle Chalk, and that the beds of chalk, formed as horizontal layers beneath the Cretaceous Seas, are now tilted down towards the SSE at an angle of about 6°. Despite efforts to clarify the nomenclature of such features there is still some confusion in the application of the term 'escarpment' which is variously used to describe the whole of the above feature or only the steep hillside (N facing in this instance). The writer therefore prefers to use the contracted form, **scarp** for the steep slope roughly at right angles to the direction of dip of the rocks, and to refer to the entire feature as a **cuesta**. The area of gradually decreasing summit heights is the **dip slope** (*Fig 157*). The width, and to some extent the height,

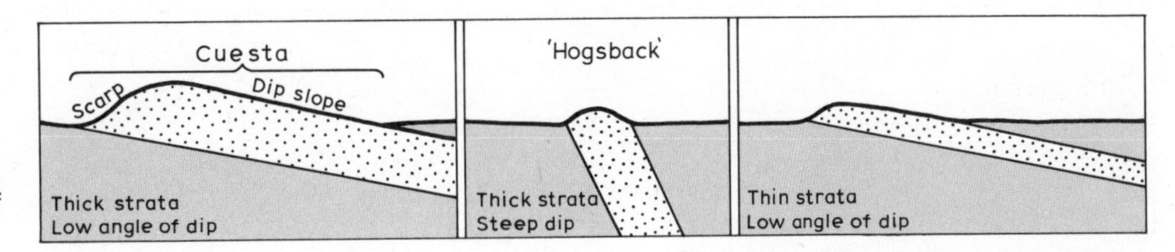

Fig 157 Sections to show relation of strata thickness and angle of dip to cuesta features

of a cuesta are related to the thickness of the outcrop concerned and its angle of dip.

It will be seen from the map extract that there are few extensive areas of flat land on the dip slope, but equally that there are no really precipitous slopes, or rocky outcrops. This characteristic of chalk landscapes has led to such descriptions as 'the whale-backed downs'; '. . . the ever-recurring double curve which was long ago styled by Hogarth *The line of Beauty*'; '. . . the surface is billowy, but not broken; the swells resemble Biscayan waves half pacified'; 'The cross-section (of a chalk valley) is usually a smooth curve'; '. . . the remarkably symmetrical double curve so characteristic of our chalk regions'. Such descriptions may be evocative and indeed poetic, but they are inaccurate. Precise measurement of slope angles and accurate surveys of chalk slopes and valley sides show that there are considerable stretches of straight slopes and that breaks of slope, while certainly not sharply etched are none the less frequently present.

Dissecting the dip-slope are clearly marked valleys, some 60 m (200 ft) lower than the interfluve ridges. The patterns of the main valleys and tributaries converging on Imber for example resemble those of river valleys, presenting a branching or **dendritic** form. The longitudinal profiles show a decrease in both altitude and gradient comparable to that found in river valleys. Transverse valley profiles, although modified in detail are broadly similar to valleys carved by river action. Yet there are no rivers in these valleys at the present time. They are **dry valleys.**

Scarp foot springs occur at about 90 m to 110 m (i.e. close to the 300 ft and 350 ft contours) in this area, and streams also appear in the valleys to the S of the map extract at about the same height, and so we are led to the conclusion that the floors of the the dry valleys which we see today lie above the general level of the water table but that at some past era the water table must have been higher to supply the rivers which eroded these valleys.

Various explanations have been suggested to account for the lowering of the water table in chalk downland:

Mode 1: Past climates, it is suggested, were much wetter than those of today and resulted in a higher water table.

Mode 2: During glacial epochs the chalk rock and subsoil was permanently frozen, rendering it impermeable so that streams flowed over the surface.

Mode 3: The gradual retreat of the scarp face due to erosion results, in the case of dipping strata, in a lowering of the junction of chalk and underlying clays. It is the height of this junction which governs the height of the water table—hence scarp recession lowers the water table (*see Fig 156*).

Many problems relating to the origins of dry valleys in chalk remain to be answered and research into the hydrology of chalk and the detailed land forms of the valleys is required. One difficulty is to determine the extent to which the valley profiles have altered since they were formed by running water. Whether or not the Pleistocene Ice Age was the prime factor in valley formation (*see Mode 2 above*), examination of the deposits on the floors of most chalk dry valleys shows that under periglacial conditions there was much slumping of saturated chalk (**solifluction**) and this has played a very significant part in the formation of the smooth, rounded appearance which is so characteristic of chalk scenery.

FURTHER WORK

Practical

(1) Construct a longitudinal profile of the Imber Valley from 929504 to Chitterne Village (993440). Note that the O.S. map indicates a stream flowing in the valley from 1 km NE of Chitterne church. This is likely to be a **winterbourne**—a stream which flows only when the water table rises a little after prolonged periods of high rainfall.

(2) Draw cross-sections to show the characteristic shape of chalk valleys, e.g. 940500 to 940470.

(3) Place tracing paper over the map extract and use the contours to reconstruct the stream pattern which once existed.

(4) If visits can be made to a chalk area, use clinometer measurements to draw an accurate section showing nature of slopes on the side of a dry valley.

Reading

Carter: *Land Forms and Life*, Section 17, pp. 136–143
*Clark, M. J.: *The Form of Chalk Slopes*, Southampton Research Series in Geog. No. 2, 1965
*Coppock, S. T.: *The Chilterns*, British Landscape through Maps series, Geog. Ass.
Dury: *The Face of the Earth*, Ch. 4, pp. 31–36
Gresswell: *Rivers and Valleys*, Ch. 3, pp. 36–38
——: *Physical Geography*, Ch. 15, pp. 131–142
Hardy & Monkhouse: *Physical Landscape in Pictures*, p. 38
Horrocks: *Physical Geography and Climatology*: Ch. 8, pp. 125–126, 133–136
Monkhouse: *Principles of Physical Geography*, Ch. 5, pp. 83 and 92–94 and 98–100
——: *Landscape from the Air*, p. 17
*Sparks: *Geomorphology*, Ch. 7, pp. 159–166
Trueman: *Geology and Scenery*, Ch. 4, pp. 43–59
Wooldridge & Morgan: *Outline of Geomorphology*, Ch. XIX, pp. 254–264

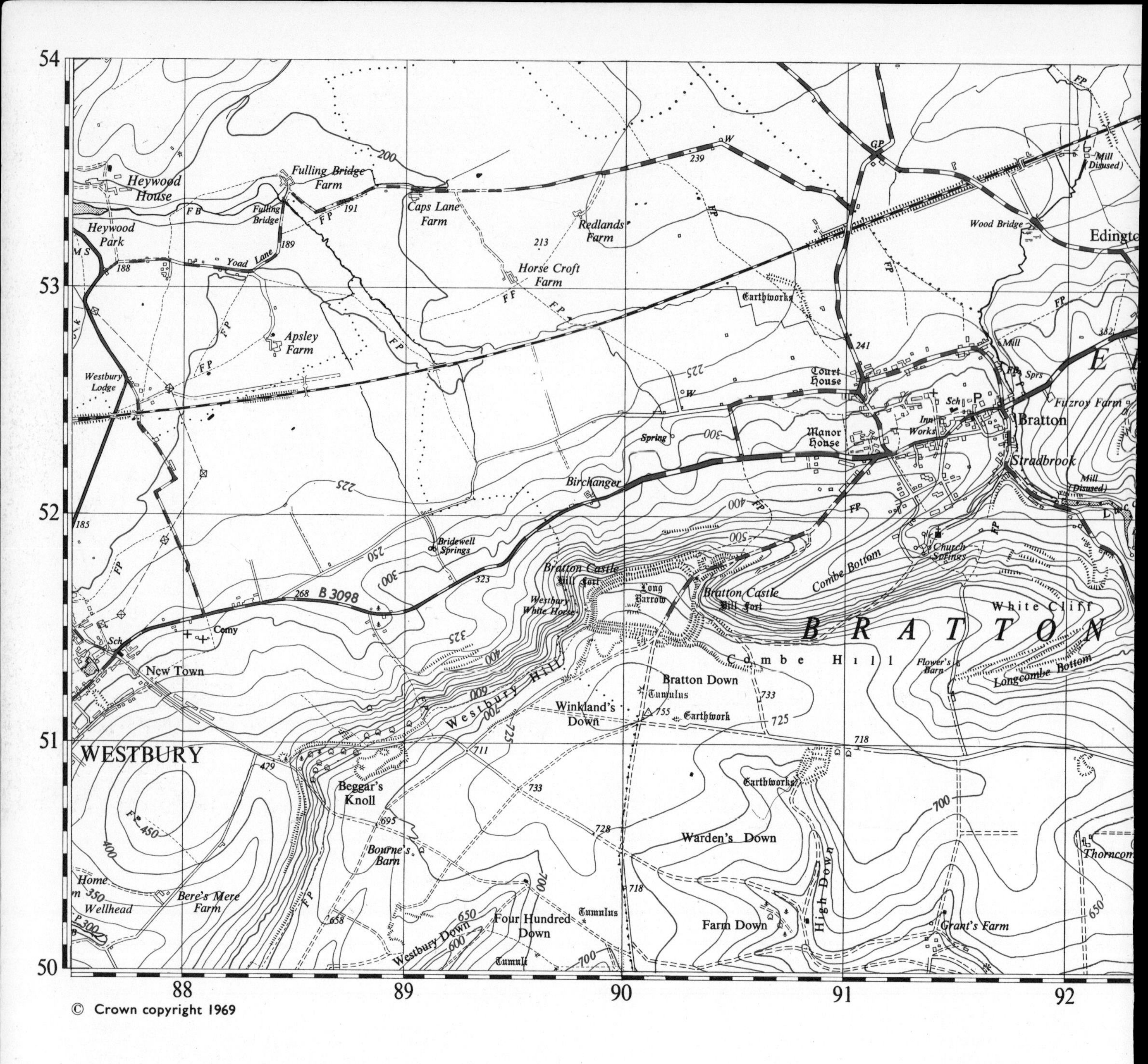

North Scarp of Salisbury Plain; Bratton, Wiltshire

Study E2

FEATURES OF A CHALK SCARP

O.S. 1 : 63 360, (1·6 cm to 1 km *or* 1 in to 1 mile)
Sheet 167, Salisbury
O.S. 1 : 25 000, (4 cm to 1 km *or* $2\frac{1}{2}$ in to 1 mile)
Sheets ST/85 and ST/95

Study A4 commenced with an investigation of a small spring at the foot of the chalk scarp at Westbury, Wilts. The portion of the scarp selected for close analysis in this section lies a short distance further east; in fact Wellheads Spring is situated about 1 km beyond the lower edge of the aerial photograph (*Fig 158*) and appears at the SW corner of the map extract above.

MAP AND PHOTOGRAPH ANALYSIS

(1) Locate on the map, the following features shown on the photograph:

(a) the road and houses situated in the left foreground of the photograph;
(b) the White Horse carved in the hillside;
(c) the railway line which can be seen cutting across the top left-hand corner of the photograph;

(2) In which direction was the camera pointing?

(3) Locate on the photograph the following features shown on the map:

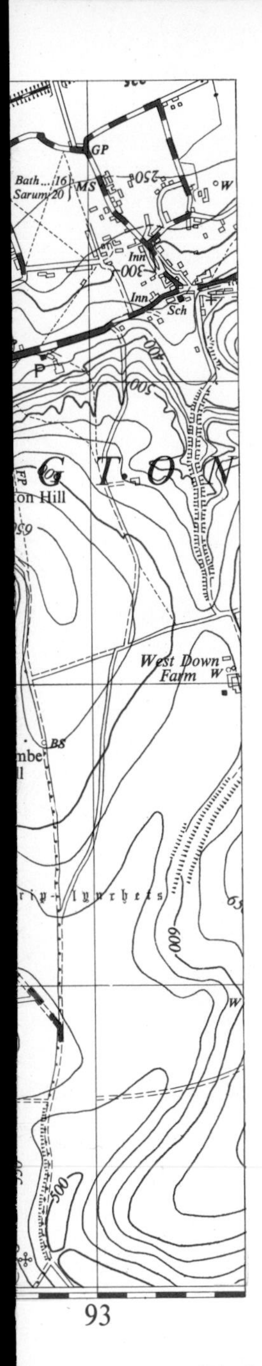

Fig 158 North scarp of Salisbury Plain, Westbury, Wiltshire

Fig 159 O.S. Bratton
Scale 1:25 000 (4 cm to 1 km *or* $2\frac{1}{2}$ in to 1 mile) Contour interval 25 ft

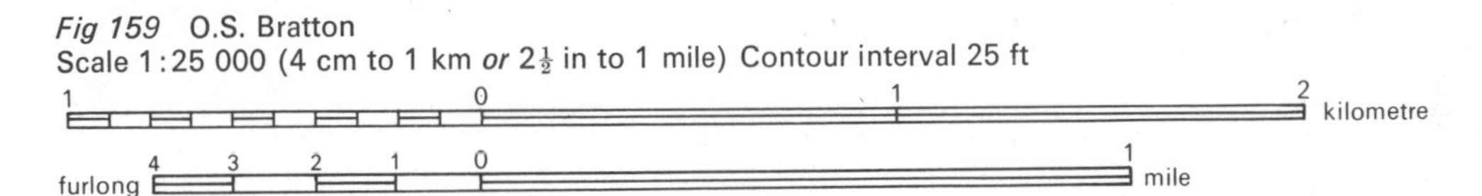

(a) Birchanger Farm—G.R. 898521;
(b) Bratton Village—G.R. 915525;
(c) Longcombe Bottom—G.R. 920514;
(d) Edington Hill—G.R. 926525.

(4) Draw a contour section from G.R. 886522 (225 ft contour) to G.R. 902500.

(5) What is the approximate gradient of the steepest part of the scarp face? Can you express this as an angle of slope, i.e. the angle between the slope and the horizontal? If necessary, draw a simple diagram and measure the angle. (Note that you will have used a degree of vertical exaggeration when drawing your section, so that this will exaggerate the angle of slope.)

(6) Draw a field sketch based on the photograph and annotate it with any names and features which you can identify from the map.

(See also suggestions for further practical work at the end of this section.)

The chalk strata of Salisbury Plain has, in the area under study, a slight regional dip towards the SSE. Study of the geological section (*Fig 160*) shows that the chalk outcrop has been truncated, and that this rock series must have extended much further N in the geological

Fig 160 Geological section through scarp west of Bratton

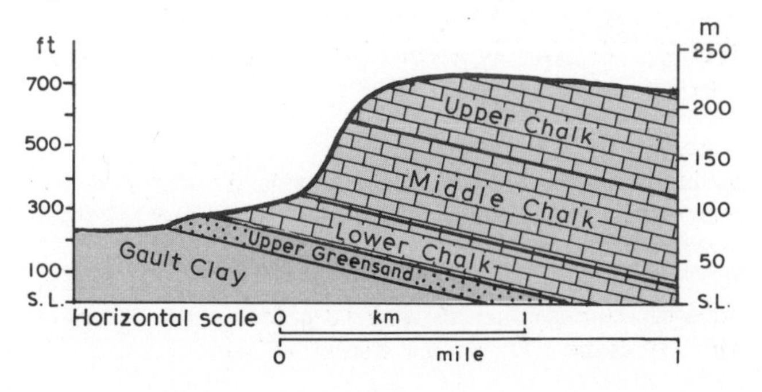

past. The mechanism of **scarp recession** poses many problems which are still imperfectly understood. In particular one might expect that if spring sapping (or headward erosion) had continued for long periods, the line of the escarpment crest would become severely indented. While such indentations exist in some places the general impression of this, and of many other scarps, is of the overall straightness of the crest line over considerable distances (cf. Further Work, Practical 1, p. *115*).

Whatever the processes which have produced it, the sweeping, turfed slopes of the scarp arc of great natural beauty and form a highly significant feature in the physical and human geography of the area.

Many of the relatively minor physical features owe their origins to recent geological history. The scarp between the White Horse and Beggar's Knoll (the wood in the centre foreground of the photograph) is marked by very clear corrugations, or fluting. These are, in some respects, reminiscent of the much more sharply etched forms known as **lapiés** in regions of karstic limestone (*see Study A1*). Possibly atmospheric and surface water solution plays some part in their origin, but their present smooth form can be attributed to **solifluction**, the slumping of masses of saturated rock under periglacial conditions. Such modification of chalk scenery has been invoked above (*p. 111*) to explain aspects of dry valley formation. The chalk of Salisbury Plain lies some 60 km south of the maximum extent of Ice Sheets during the Pleistocene period. Clearly the region under study must then have been a region of intense cold with tundra-like conditions; snow covered it for the greater part of the year, with any water in the rock frozen to very considerable depths—**Permafrost**. During brief, cool summer periods, only a thin surface layer could be thawed, with the underlying Permafrost remaining to convert the chalk into an impervious mass. Under such conditions the saturated surface layers became so lubricated in summer by melt water that 'mud glaciers' could begin to slide down any steep slope. The debris from such mass slumping (**solifluction**) has since been spread over the scarp foot zone where its presence is attested by the numerous flint fragments, bleached by processes which operate under conditions of intense cold, and found widely on strata which do not themselves contain flint.

Your contour section (Qu. 4 above) will show a platform almost a kilometre wide at the foot of the scarp, at a level of about 90 m (300 ft contour). This corresponds fairly closely with the outcrop of the Lower Chalk (*Fig 160*). Detailed research in other parts of Southern England suggests that this feature is not entirely structural in origin, but that it represents a portion of a previously widespread erosion surface related to a time during the post-glacial period when sea-level was some 55 m (180 ft) above its present level. All river valleys and plains would then have been graded to this higher sea-level, with of course increasing altitude with greater distance from the sea. Thus, just as the valley of the Bristol Avon to the N of the map extract is today about 36 m (120 ft) above sea-level, the scarp foot platform lies 36 m above the sea-level postulated at the time of its formation.

A combination of scarp recession, rejuvenation and periglacial modification seems to be involved in an explanation of the striking scarp-combes near Bratton. These are three steep-sided and partially dry valleys. Longcombe Bottom (920514) can be easily distinguished on the map and in the background, right of centre, on the photograph. Luccombe Bottom (926523) runs from Edington Hill towards White Cliff, and Combe Bottom (910518) lies between the minor road to Bratton Castle

Fig 161 Stages in the formation of scarp combes

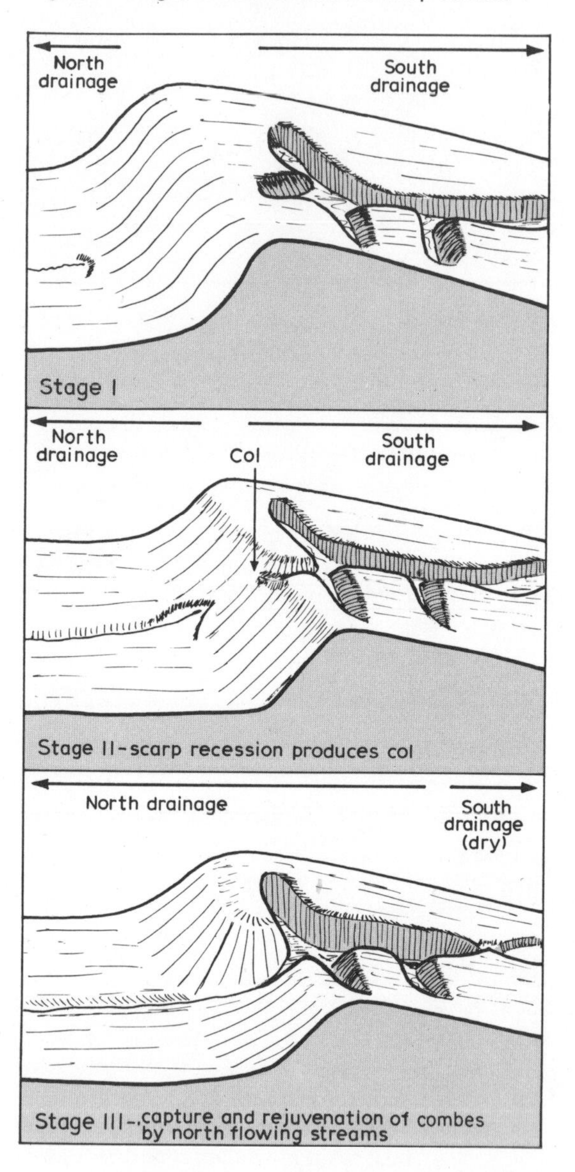

and Combe Hill. It is suggested (*R. J. Small—reading list at end of Study*) that these valleys once formed part of the dry valley system running S and E towards Imber. As the scarp receded, cutting back into this system, a col was exposed. With the rejuvenation of drainage particularly affecting streams flowing N towards the Bristol Avon, headward erosion by such streams enabled them to capture the three combes—the N ones being affected first. Thus their drainage was diverted from S to N, and the combes themselves were deepened and enlarged (*Fig 161*). It will be noticed that as the scarp has receded, the N side of Combe Bottom has been partially removed, so that now only a low spur eastward from Bratton Castle separates this valley from the scarp foot zone.

The preceding summary of a probable origin of the Bratton scarp combes illustrates well the complexities which must be unravelled in order to understand the formation of present-day landscape features. Hypotheses of this nature require detailed observation and measurement in the field, as well as a clear understanding of the processes and structures involved. Only when much detailed study of actual examples has been made can the research worker suggest that his general line of thought may be usefully applied to similar features elsewhere.

FURTHER WORK

Practical

(1) Map study should be undertaken of other chalk escarpments, and related photographs if possible. The student should particularly seek similarities to, and differences from, the area studied above. Maps particularly suitable are: O.S. 1:63 360, Sheet 182, Brighton and Worthing. (Note that a similar origin for Devil's Dyke (G.R. 263110) has been postulated, as that for the Bratton combes); a contrasting chalk scarp with deep re-entrant combes can be seen at Batcombe (620040) on O.S. 1:63 360, Sheet 178, Dorchester. See also Sheet 93 (scarp of Yorkshire Wolds) and Sheet 159 (scarp of Chilterns).

(2) Although this book is not concerned with aspects of human geography, many students will wish to relate the physical features of chalk escarpments to such aspects as land use, prehistoric occupation, settlement sites and communications. In this context, the author would point out that many easy generalizations on this subject are not always borne out by detailed study. For example, how many so-called 'spring line' villages are actually sited at a spring?

Reading

Holmes: *Principles of Physical Geology*, Ch. XIX, pp. 566–569

Horrocks: *Physical Geography and Climatology*, Ch. 8, pp. 133–136; Ch. 5, pp. 75–78

Monkhouse: *Principles of Physical Geography*, Ch. 5, pp. 84–87 and 98–100

——: *Landscape from the Air*, p. 1

*Sparks: *Geomorphology*, Ch. 7, pp. 159–166

*R. J. Small: *Role of Spring Sapping in the formation of Chalk Escarpment Valleys*, 'Southampton Research Series in Geog.,' No. 1, 1965

*——: *Escarpment Dry Valleys of the Wiltshire Chalk*, Institute of Brit. Geog.; Trans. and Papers, No. 34, 1964

Trueman: *Geology and Scenery* Ch. 4, pp. 42–59

Treak Cliff Area, Castleton, Derbyshire

Study E3

UNDERGROUND DRAINAGE IN CARBONIFEROUS LIMESTONE

O.S. 1:63 360, (1·6 cm to 1 km *or* 1 in to 1 mile) Tourist Edition, Peak District *or* O.S. 1:63 360, Sheet 111, Buxton and Matlock
O.S. 1:25 000, (4 cm to 1 km *or* 2½ in to 1 mile) Sheet SK 18

South of the line marked approximately by the A625 from Hathersage and Castleton to Chapel-en-le-Frith (*Fig 162*), the Carboniferous Limestone plateau of South Derbyshire stretches S for over 30 km to the mouth of Dove Dale. Travelling westward along this road from Castleton, one first traverses a kilometre of the gently sloping land which marks the W end of the Hope (or Castleton) Valley. To the S the land rises steeply a short distance from the road to a skyline some 100 m (contours 600 ft in valley, to 900 ft on hillside) above road level, and then as the map shows, continues to an altitude of 400 metres (1300 ft contour) and eventually to 450 m (1450 ft contour). Barring the route westward is the steep (30°) slope of Treak Cliff rising 180 m above the level of the road (700 ft to 1300 ft contours), and a little to the NW is the equally steep ridge of Cold Side, culminating in the almost sheer triangular cliff which marks the E face of Mam Tor, reaching an altitude of 500 m (1600 ft contour) and showing exposures of horizontally bedded sandstones and shales. To avoid these two obstacles, the road is engineered NNW along the foot of Treak Cliff and then, climbing at a steep gradient, turns back upon itself in a hairpin bend to cross the lower slope of Mam Tor. This section of the road is in constant need of maintenance because of the unstable nature of the rock; waterlogged sandstone over shale is continually slumping and thus

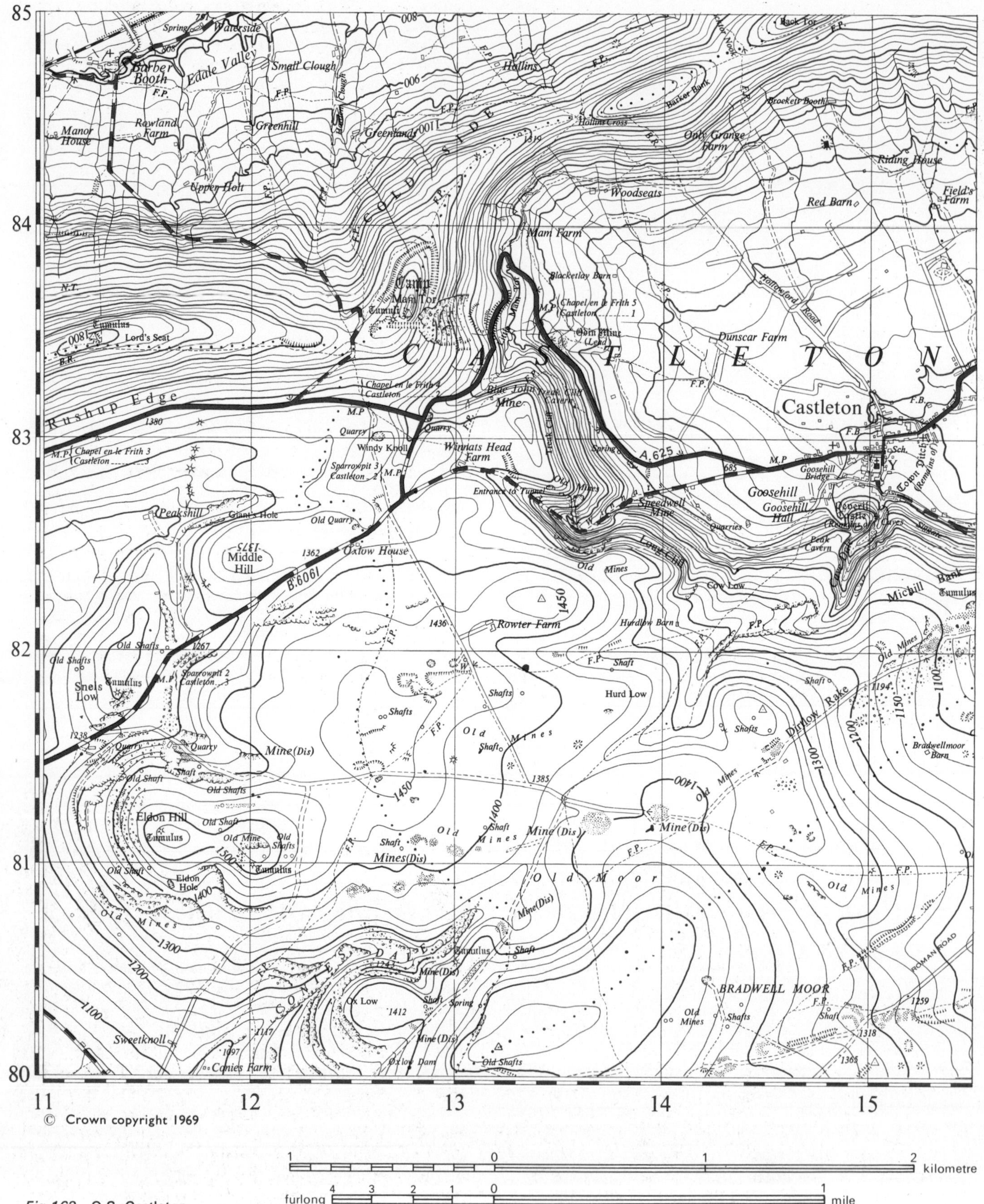

Fig 162 O.S. Castleton
Scale 1:25 000 (4 cm to 1 km *or* 2½ in to 1 mile)
Contour interval 25 ft

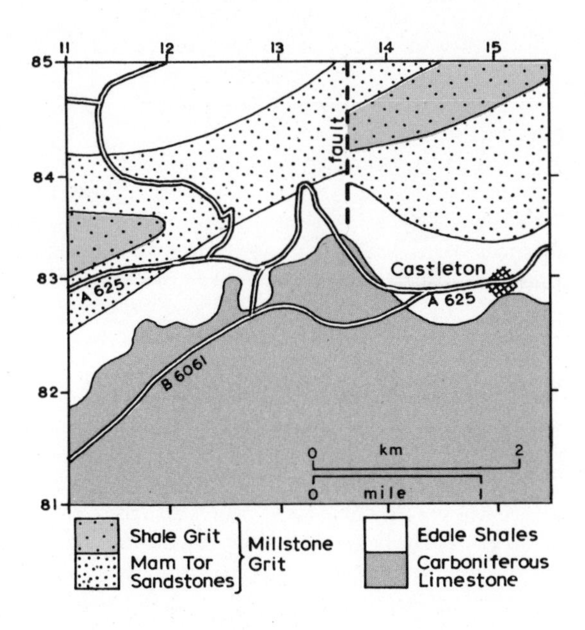

Fig 163 Treak Cliff Area, geological map of area shown on

undermining the road's foundations. For this reason Mam Tor is locally known as the Shivering Mountain. Finally the road utilizes the side of a small stream valley to climb to 400 m (1300 ft contour) and thence follow the S side of Rushup Edge westward.

Looking S from A625 as it follows the line of Rushup Edge, one may see a number of small streams mostly rising near Rushup Edge and flowing S across a broad, shallow valley. Each stream has carved a narrow gulley which comes to an abrupt end at the S edge of the valley (e.g. 112822). The significant point about these streams is that on the map they seem to come to an end without joining a larger river. If this were an arid region one could explain such a feature in terms of temporary streams and high rates of evaporation, but this makes no sense anywhere in the British Isles, and certainly cannot be a reasonable explanation in an area of high rainfall such as the Pennines. Since these streams do not dry up, they can only disappear from the map because their water passes undergound.

EXERCISE

Place a piece of tracing paper over the map extract (*Fig 162*) and use the geological map (*Fig 163*) to draw in the rock outcrops on the tracing paper. Trace the course of streams and use a symbol to indicate the position of dry valleys. What conclusion do you reach regarding the characteristics of Carboniferous Limestone?

An alternative route W from Castleton leads, by way of a minor road, through the Winnats Pass (138826). This is an extremely narrow and steep-sided dry valley. So steep, in fact, are the sides rising 125 m above the road (900 ft to 1300 ft contours) that the term **gorge** could be correctly applied to this feature (*Fig 164*). The lower slopes are, for the most part, grass covered scree slopes of up to 40°, and above these tower near vertical cliffs of bare grey to white limestone, fretted by weathering so that in places there are narrow rock buttresses crowned by pinnacles jutting into the line of the gorge. At one or two points on the valley sides, clefts in the rock have been enlarged to form cave-like features. The road westward

Fig 164 Winnats Pass

through the Winnats Pass climbs steeply with gradients of up to 1 in 4, and finally emerges at 410 m (spot height 1362 at G.R. 123824) on to more gently sloping undulating land near the summit of the limestone plateau. The upland is mainly of gentle slope but blocks of bare limestone jut through the surface, and here and there grassy or rocky hollows occur.

Within the triangle formed by the Winnats Pass and the hairpin bend of the A625 are three sets of caves which are open to the public: The Blue John Mine (132832); Treak Cliff Caverns (136831) and Speedwell Mine (139827). In each case the caverns which are shown to visitors are partially natural in origin and partially artificial. The man-hewn caves and tunnels are the result of relatively small-scale mining for two substances—lead ore (*Galena*) which is commonly found in veins in the Carboniferous Limestone of Derbyshire and elsewhere; and Blue John, a unique form of fluorspar tinted purple by the presence of hydrocarbons, which is found nowhere outside the small area of Treak Cliff. As the miners of past centuries tunnelled their way into Treak Cliff, they broke through into systems of natural caves frequently beautified by displays of stalactites, stalagmites and wall coverings of tinted calcite resembling frozen waterfalls. Inside these natural caves are large masses of clayey debris which can be shown to have been deposited by underground rivers, and if further proof is required that the caves were the result of erosion by running water, it can be found in swirl marks and potholes[1] exactly similar to those found in the bed of a river (*Study A5*), but in this case found in the roof of the caves which must, therefore, have been completely filled with water, probably under some pressure. In Speedwell Mine the underground stream remains, and visitors are taken along a portion of it by boat. (Notice that Speedwell is at a lower elevation than the other two cave systems and thus reaches to below the present water table in the limestone.)

There are certain aspects of the preceding description of the Carboniferous Limestone upland of North Derbyshire which may seem to parallel features which were noted for the chalk downland of Wiltshire, e.g. the lack of surface drainage and the occurrence of dry valleys. And, of course, the chemical composition of the bedrock is the same in both areas, namely calcium carbonate, so that both are subject to solution in rain water (*p. 3*). Closer examination of the map, however, reveals many points of difference, and these differences are so readily apparent in the field that the landscapes in fact present very contrasting 'personalities'. Some of the more significant of these differences are:

(a) Altitude: while chalkland summits average 180–250 m and only occasionally exceed 270 m, the hilltops throughout the Carboniferous Limestone area of Derbyshire exceed 300 m and, as the map extract shows, rise to 450 m in the Castleton area.

[1] The student should not confuse the small circular holes drilled, usually in the bed of a river, and forming part of the process of vertical erosion (*p. 14*), with systems of underground caves in limestone which are also often called 'potholes'.

(b) Nature of Summits: the rounded and relatively narrow summit 'ridges' and spurs of chalk scenery, dissected by a well developed dendritic pattern of dry valleys, is replaced by a plateau-like upland surface notched at its edges by deep dales frequently lacking any indication of tributary valleys.

(c) The upland surface of the Carboniferous Limestone is pockmarked by hollows, some only a few metres across, others with a diameter of 100 m or more. While some of these hollows may be attributed to the activities of miners in past centuries, others are of natural formation due to underground solution and only rarely do comparable features occur in chalk areas.[1]

(d) One of the most striking characteristics of Carboniferous Limestone areas in the field is the frequency of the occurrence of outcrops of bare rock. This has been noted above as one of the features of the Winnats Pass, but over the whole area lines of bare rocks (**scars**) and isolated boulders of limestone protrude from the turf. This characteristic is given added emphasis by the fact that most field boundaries are of dry-stone walling of the same greyish rock.

(e) Occasionally a river which crosses an outcrop of chalk, such as the Mole N of Dorking, loses a proportion of its water by underground seepage and in an exceptionally dry season may disappear entirely. But this is quite rare, whereas it is very common in areas of Carboniferous Limestone to find streams which plunge or seep beneath the surface, sometimes down nearly vertical clefts in the rock.

(f) Streams which disappear in this way flow, often for considerable distances, through a network of underground channels. The caves thus formed may, at a later stage, be abandoned as have the higher cave systems of Treak Cliff. Natural caves are rarely found in chalk.

These differences may to a large extent be explained by the differing physical characteristics of chalk and Carboniferous Limestone. The latter is much stronger and harder, as can easily be determined from a comparison of specimens of the two rocks. The system of jointing is also much more clearly developed in Carboniferous Limestone to form a regular pattern of horizontal and vertical joints dividing the rock into nearly rectangular blocks; in chalk on the other hand, the joints form a very irregular pattern. It is the joints in limestone of all types which are most easily affected by the solvent action of rain water and which offer the easiest route for water to pass through the rock. Where the joint system offers a continuous passage for underground water, the steady flow which ensues provides a constant supply of percolating water to further enlarge the joints, and this eventually forms a well developed cave system through

[1] In this connection it may be noted that some depressions do occur in chalk which has had a thin overlay of Tertiary gravels (e.g. South Mimms, Hertfordshire) and in places where such a thin deposit still covers the chalk (e.g. Puddletown and Affpuddle Heaths, Dorset). Solution hollows of this type may be attributed to the downward movement of acidic ground water which accumulates in the overlying gravels (*see B. W. Sparks, Geomorphology p.* 160).

which sizeable streams may flow. In chalk, lacking a continuous set of joints, the rate of water movement underground is much slower and the amount of solution consequently less. Also, since chalk is less strong, many fissures which are opened by solution tend to be closed up by the weight of fractured chalk above.

Water is only able to pass downwards through Carboniferous Limestone at levels above the water table. Once the water table is reached the movement will be lateral, dependent upon the creation of a 'head' of water produced by a doming up of the water table beneath the highest ground. Once the joints have been sufficiently enlarged by solution the underground water can flow freely through cave systems and in this case the water table may be almost flat. The greatest lateral flow is likely to occur near the upper level of saturation, and it is here that the greatest development of caves is to be expected. Any subsequent lowering of the water table will result in such caves being left permanently dry, as for example those of Treak Cliff Caverns.

Although Carboniferous Limestone is sufficiently strong to sustain very large underground caverns, there may come a time when solution below the surface and denudation at the surface combine to leave only a very thin roof layer. This may then collapse either as a consequence of its own instability or as a result of earth tremors. Such a roof collapse along a line of caves would produce a gorge, perhaps with remnants forming natural bridges, and it has been suggested that the Winnats Pass may have been formed in this way. A similar origin has been postulated for Cheddar Gorge, but it should be noted that some experts consider that the collapse of cavern roofs to form gorges is a rare occurrence and that most examples of gorges in limestone areas must be explained by other processes.

The landforms described above for the Castleton area may be said to be representative of the scenery developed on similar limestones both in other areas of Britain and overseas. In the case of North Derbyshire the limestone is of Carboniferous age; in the Causses of the French Massif Central and in the Karst district of Yugoslavia the bedrock is limestone of the Mesozoic era. It is from the extensive development of limestone features in the Yugoslav region that this scenic type is known as **Karst scenery**.

FURTHER WORK

See end of next Study.

Reading

K. C. Edwards: *The Peak District*, Ch. 3, pp. 28–49. Collins 1962

Gresswell: *Geology for Geographers*, Ch. 11, pp. 122–123

——: *Physical Geography*, Ch. 15, pp. 132–138, 142–145

Horrocks: *Physical Geography and Climatology*, Ch. 8, pp. 126–133

Monkhouse: *Principles of Physical Geography*, Ch. 5, pp. 94–97, 100–101, 103–104

Scovel: *Atlas of Landforms*, pp. 76–83

Sparks: *Geomorphology*, Ch. 7, pp. 154–159

Strahler: *Physical Geography*, Ch. 29, pp. 459–461; Ex. 3, p. 469

Trueman: *Geology and Scenery*, Ch. 11, pp. 142–147, 151–156; Ch. 12, pp. 157–160

Wooldridge & Morgan: *An Outline of Geomorphology*, Ch. XIX, pp. 264–270

Geologists' Association Guides: *The Castleton and Edale District*

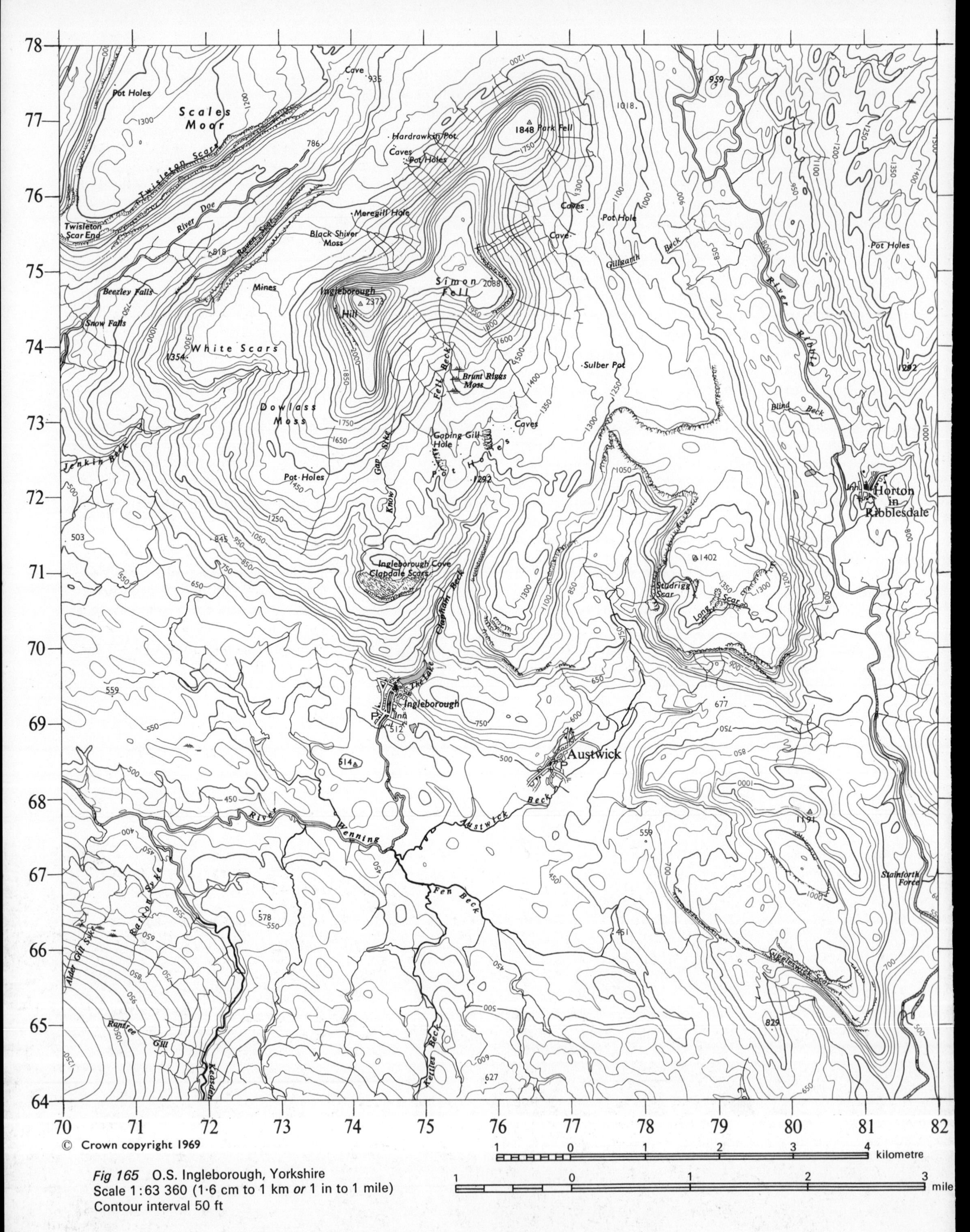

© Crown copyright 1969

Fig 165 O.S. Ingleborough, Yorkshire
Scale 1:63 360 (1·6 cm to 1 km *or* 1 in to 1 mile)
Contour interval 50 ft

LIMESTONE UPLAND AND FAULT SCARP

O.S. 1:63 360 (1·6 cm to 1 km *or* 1 in to 1 mile)
Sheet 90, Wensleydale

MAP AND PHOTOGRAPH INTERPRETATION

(1) Draw a contour section from 710760 to 810715. Annotate your section to show Rivers Doe and Ribble, Ingleborough, Raven Scar and other scars.

(2) Follow all the streams which rise round the summit of Ingleborough, determine the heights at which the streams disappear and draw up a list showing the number of streams disappearing within each 50 ft (151 m) height range from 1100 ft (335 m) to 1400 ft (335–425 m). What percentage of disappearances occur (a) between 1300 ft (400 m) and 1400 ft (425 m)? (b) between 1200 ft (365 m) and 1300 ft (400 m)? What is the level below which streams reappear? What does this information suggest concerning the nature of the rock between 800 ft (245 m) and 1200 ft (365 m)? *Note:* one stream flows continuously from its source at 1800 ft (520 m) to below 500 ft (150 m). What evidence is there that the course of this stream is partly artificial?

(3) Study the photograph of Gaping Gill (*Fig* 166). Where does the stream flow after rain? Estimate the difference in level between the mouth of the cave and the general land surface. Write a short description of this feature. Locate Gaping Gill on the map extract: how can you explain the line of potholes to the S of Gaping Gill continuing the line of the stream?

(4) Place a sheet of tracing paper over the map and indicate by suitable symbols all evidence of limestone scenery—potholes, disappearing streams, caves, dry valleys, scars (linear outcrops of bare rock). *Note* that scars can occur on other rocks, and that the scar near G.R. 760750 is not a limestone feature. Using this tracing in conjunction with a study of the drainage pattern it is possible to sketch in those areas in which it is likely that limestone forms the surface rock. Compare your result with the geological map (*Fig 168*).

(5) How do Clapdale Scars (G.R. 743709) differ from other scars shown in this area? Limestone slopes are normally covered by a thin soil and short grass, but occasionally almost horizontal areas of bare limestone are exposed. When this occurs the joint system is subjected to solution; as the joints are widened the area is carved into a series of blocks separated by widened cracks up to 1 ft across and often at least as deep. (*Fig 167*). In Yorkshire these blocks are known as **clints** and the enlarged joints as **grikes**, and the whole feature is a **limestone pavement.** It may be noted that some experts consider that the limestone pavements occurring in Northern England may have originally been formed beneath a cover of vegetation and have recently been exposed following climatic changes.

(6) Draw a contour section from 803660 to 800653. Compare your section with the photograph of Giggleswick Scar (*Fig 170*). Use map and photograph to estimate the angle of the slope just in front of the wooded hillside in the foreground. Where does the slope appear to be steeper than this? Use map and photograph to list the differences between the land to the NE and SW of the

Fig 166 Gaping Gill

Fig 167 Limestone pavement above Malham Cove, Yorks

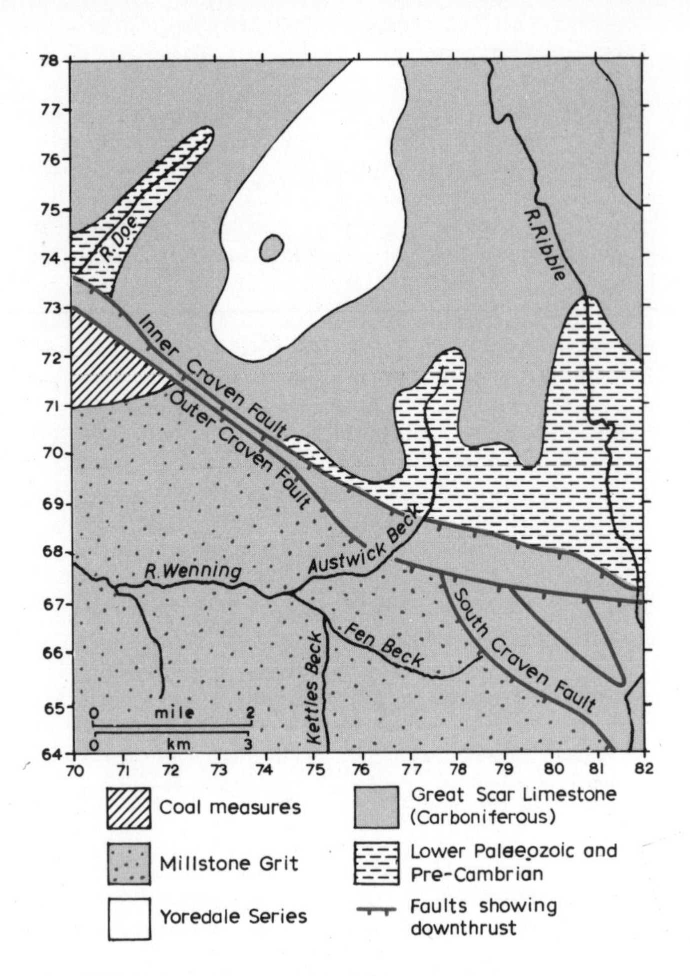

Fig 168a Geological map of Ingleborough area

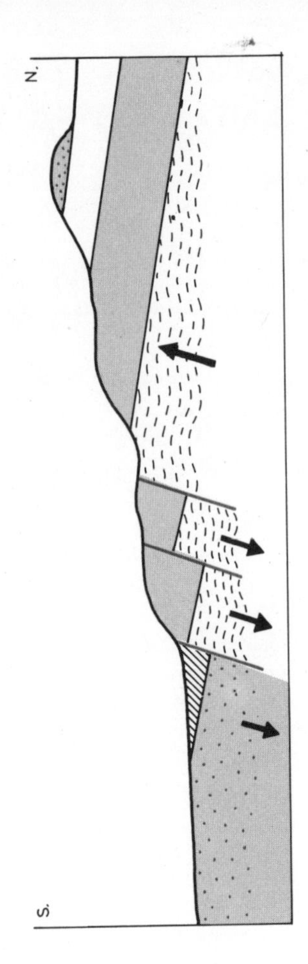

Fig 168b Generalized N-S section to show relationship of strata to relief

steep slope. The geological map (*Fig 168a*) shows the land to the SW is Millstone Grit and to the NE is Carboniferous (Great Scar) Limestone. But Millstone Grit is a younger rock than Carboniferous Limestone; how then does it come to occupy the lower ground?

The answer to the latter problem lies in the fact that Giggleswick Scar is a **fault line scarp.** This fault had been active in Carboniferous times and then at a period post-dating the deposition of Carboniferous rocks (in fact during Tertiary times) the land to the NW of this fault (the South Craven Fault) was uplifted relative to the land to the S by over 1500 m (5000 ft). It might of course be equally true to say that the land SW of the fault was lowered relative to that to the NE by over 1500 m. This enormous movement did not occur overnight but was the total result of thousands of small shifts spread perhaps through millions of years. As upthrusting occurred, the higher, north eastern, side was exposed to greater erosion so that in total far more erosion has taken place on this side o the fault than on the downthrow side. Thus although ne fault line scarp of today preserves the up throw side a s the higher ground, the difference in height is much less than the vertical displacement of the rocks (*Fig 169*), and it will be realized that continued denudation may remove the surface expression of the underlying fault. It can be seen, for example, that the Inner or North Craven Fault is not marked by any clear-cut scarp.[1]

The surface features of this area cannot be understood without reference to its geology and to its structural history. The foundation of the area is of highly folded pre-Cambrian and lower Palaeozoic rocks. Upon this basement, after it had been reduced to a peneplain, rocks of Carboniferous age were deposited: first the Great Scar Limestone, very pure and well bedded; then the shales, limestones and sandstones in alternating bands which form the Yoredale Series, overlain in turn by the Millstone Grit and Coal Measures. At the close of the Palaeozoic era, movement occurred along the Inner Craven Fault and then in Tertiary times came the great upheaval along the outer Craven Fault described above. Erosion on the up throw side of the faults has removed almost all the Millstone Grit except for a small capping on the summit of Ingleborough, and the rivers Doe and Ribble and the Austwick Beck have cut down to expose the basement of Lower Palaeozoic and pre-Cambrian rocks. These impermeable rocks check the downward percolation of water which occurs within the Great Scar Limestone and their outcrop is marked by the reappearance of streams which disappeared at a higher level. To the SW of the faults Millstone Grit is preserved as the

[1] A fault scarp is a feature, the origin of which is directly due to the displacement of rocks by faulting.

A fault-line scarp is a steep slope developed along a line of faulting by the differential erosion of rocks on either side of the fault.

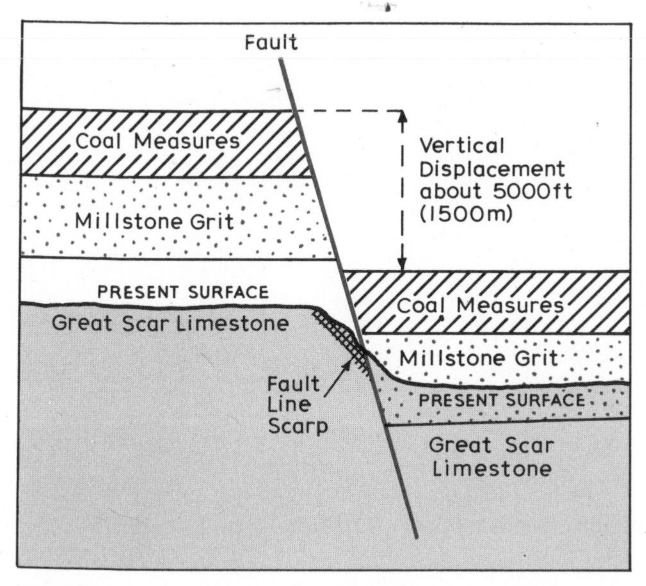

Fig 169 Section through Giggleswick Fault Line Scarp (after Gresswell)

surface rock with a very small area of Coal Measures at the W edge of the map.

Finally the landscape of the whole area has been modified by glaciation. Some 70 000 years ago only the highest summits stood out from the ice, much weathered material was removed, and a tendency for U-shaped cross-sections in some valleys can be noticed. Today the most obvious evidence of glaciation is to be seen in swarms of drumlins which are to be found in upper Ribblesdale and in the valley of the R. Wenning in the S.

FURTHER WORK

Practical

(1) Use O.S. 1:63 360, Sheet 90, in conjunction with King *Yorkshire Dales*, Monkhouse, pp. 96–97 and Plate XXVI; Educational Productions Ltd. Filmstrip *Malham*; and/or 1:63 360 Geological map to draw a carefully annotated sketch map and sections explaining the features of the area between Malham Tarn and Malham Village.

(2) Use O.S. 1:63 360, Sheet 165, or O.S. 1:25 000, Sheet ST 45, to compare the features of Cheddar Gorge with Winnats Pass. On the same maps notice the disappearing streams at 467584, 475584, etc. and ascertain the geological conditions leading to such disappearance.

Reading

Carter: *Land Forms and Life*, Section 14, pp. 105–118
Dury: *The Face of the Earth*, Ch. 4, pp. 36–39; Faulting, Ch. 5, pp. 57–60
Gresswell: *Rivers and Valleys*, Ch. 3, pp. 32–36
Hardy & Monkhouse: *Physical Landscape in Pictures*, pp. 19, 35–37
Holmes: *Princ. of Phys. Geol.*, Ch. XV, pp. 421–425
C. A. M. King: *Yorkshire Dales*, 'British Landscape through Maps' Series, Geog. Ass.
Monkhouse: *Principles of Physical Geography*, Ch. 5, pp. 94–97, 100–101, 103–104
——: *Landscape from the Air*, pp. 4, 7, 15, 16
Scovel: *Atlas of Landforms*, pp. 42–51
Sparks: *Geomorphology*, Ch. 7, pp. 154–159
Strahler: *Phys. Geog.*, Ch. 30, pp. 474–480; Ex. 2, p. 483
*M. M. Sweeting: 'Erosion Cycles and Limestone Caverns in the Ingleborough District', *Geog. Jour.*, vol. 115, 1950
Trueman: *Geology and Scenery*, Ch. 11, pp. 147–150

Fig 170 Faultline scarp, Giggleswick Scar, Yorkshire

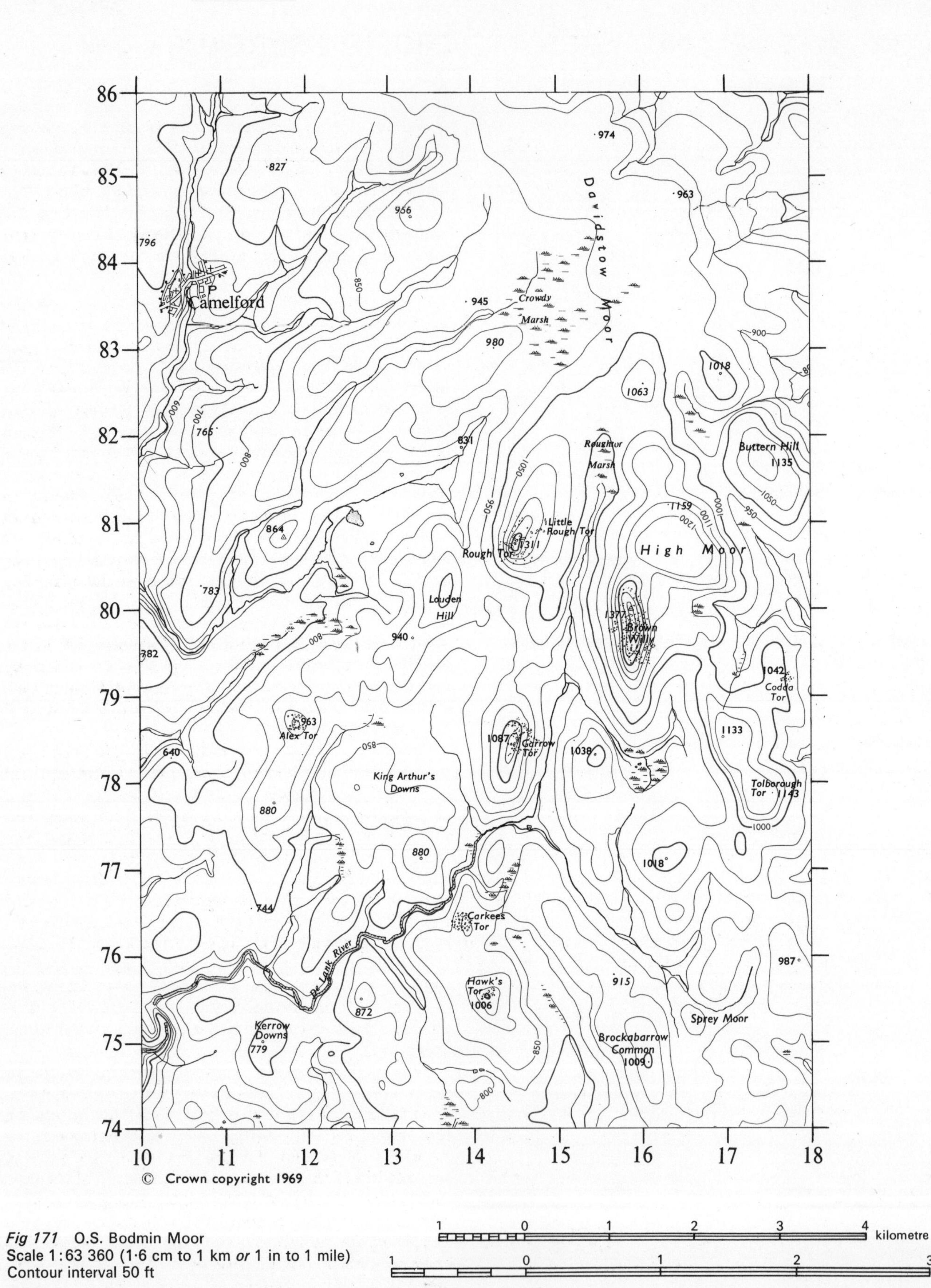

Fig 171 O.S. Bodmin Moor
Scale 1:63 360 (1·6 cm to 1 km *or* 1 in to 1 mile)
Contour interval 50 ft

1 0 1 2 3 4 kilometre

1 0 1 2 3 mile

Bodmin Moor, Cornwall

Study E5

A GRANITE UPLAND; TORS AND EROSION SURFACES

O.S. 1:63 360 (1·6 cm to 1 km *or* 1 in to 1 mile)
Sheet 186, Bodmin and Launceston

The map extract shows an area in the N of Bodmin Moor in which the landforms are consequent in part upon the nature of the bedrock, and in part upon the geomorphological evolution of the region.

The following exercises are designed to produce a descriptive account of the landforms of the area.

EXERCISES

(1) What is (a) the highest, (b) the lowest point in the area?

(2) Roughly what proportion of the map extract lies (a) above 300 m (1000 ft), (b) below 230 m (750 ft)?

(3) A 20° slope is represented by contour spacing equivalent to five contours in $\frac{1}{10}$ in (or 2 contours in 1 mm). How many slopes can you locate which are as steep as this?

(4) Describe the pattern of drainage of the area shown on the map extract. Is the stream network a close one (e.g. what proportion of the Kilometre Grid Squares have no indication of surface water)? What drainage characteristic is common to the heads of many valleys?

(5) Draw cross-sections of the stream valley between Rough Tor (145808) and Brown Willy (159800), and of the valley in square 1079. Compare the characteristics of these two valleys.

(6) Draw a projected profile of the area, viewed from the W.

Method: (a) Draw N–S lines across the map at regular intervals, e.g. through the summit of Brown Willy and then at 2 cm intervals E and W of this. Alternatively the grid lines may be used.

(b) Construct contour sections along each line, using a constant vertical scale.

(c) Trace the most westerly contour section.

(d) Place this tracing on the next contour section and add only those parts of the second section which show above the first section.

(e) Repeat for each section drawn. The final result approximates to a view of the area from the W and may suggest heights at which extensive areas of level ground occur (*see Fig 172 for example*).

Fig 172 A projected profile. There is a clear indication on this diagram of an accordant summit level just over 150 m (500 ft) and suggestion of a 115 m (350 ft) erosion surface

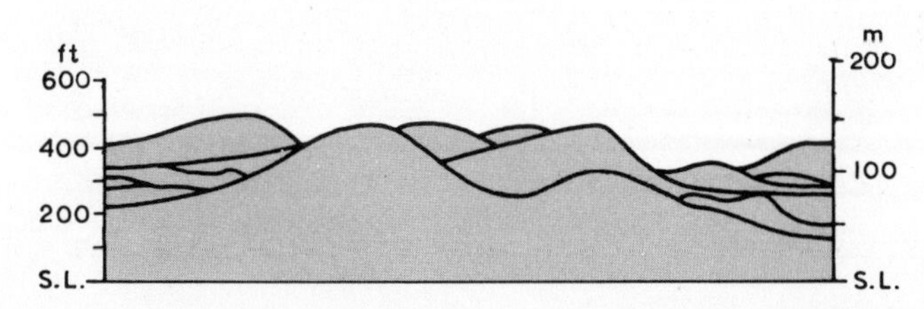

(7) Use all the information gained from the preceding exercises to write a description of the landforms of the area as shown on the map. *Note*: Such a *description* should precede any explanation or interpretation.

(8) Study the photographs of Rough Tor (*Fig 173* and *Fig 174*), and add to your description any facts revealed by the pictures which were not apparent from the map.

Bodmin Moor is one of the exposures of granitic domes intruded into the Devonian and Carboniferous rocks of the SW peninsula. Through the succeeding geological epochs denudation slowly removed rocks to expose the granitic masses which, because of their greater resistance, remained as upland areas surrounded by less elevated regions on Devonian and Carboniferous rocks. In Miocene or Pliocene times (the dating is uncertain) only the highest points protruded above sea level to form islands comparable to the Isles of Scilly at the present time. Since then sea-level has fallen in stages, with occasional reversals of this trend during inter- and post-glacial times. At heights at which sea-level remained constant for long periods one may expect to find relatively flat planation surfaces produced either by marine erosion (wave cut platforms) or by fluvial erosion (coastal plains and the lower portions of river valleys). After each fall in sea level, existing surfaces were subject to dissection by rejuvenated streams, so that only small portions of each erosion surface can be expected to remain as features of the present landscape.

Granite is an igneous, crystalline rock of great physical strength and resistance to erosion. It is to be expected then that it will stand out as the highest ground in a region otherwise composed of sedimentary rock, and Brown Willy is in fact the highest point in Cornwall. The resistance of granite to weathering results in there being only a thin, gritty soil cover—note the boulders protruding through the turf in the foreground of *Fig 173*—and this soil is generally infertile as suggested by the prevalence of the symbol for rough grazing on the O.S. map. Granite is a non-porous rock; and though it may be rendered pervious by the percolation down joints, the frequency of marshy hollows at the heads of valleys indicates the limited downward movement of water which is possible.

Tors are characteristic features of granite landscapes in south-west England, though they are by no means confined to this area nor to this type of rock. A tor consists of a mass of bare rock protruding up to 15 m through the surface at the summit of a hill, or on a hillside. Some of the boulders of the mass are an attached part of the bedrock; others are merely resting on the top. The origin of tors has attracted considerable attention from geomorphologists in recent years. Professor Linton has explained their origin with reference to the well-developed jointing system of irregular spacing which exists in the granites of, for example, Bodmin Moor (*Fig 175 I*). Under warm, humid conditions such as existed in interglacial periods the granite was subjected to decomposition beneath the surface, with the most rapid weathering occurring along joint planes. Where the distance between joint planes was largest, masses of granite remained unweathered (*Fig 175 II*). Subsequent denudation, perhaps under periglacial conditions, has removed the residue of weathering, leaving the unweathered blocks standing out as tors (*Fig 175 III*). Other theories of tor formation have been advanced and further research on this subject is proceeding.

Thus certain physical characteristics of the landscape can be directly related to the nature of the bedrock. Other characteristics, which are studied below, are the consequences of the erosional history of the area.

Fig 173 Close up of Rough Tor, Bodmin Moor

A projected profile of the area shown in the map extract illustrates the fact that this portion of Bodmin Moor consists of broad areas of nearly level surface above which rise the higher hill tops such as Rough Tor and Brown Willy. It is not difficult to visualize these protuberances jutting out to form islands above a sea-level at about 300 m above the present Ordnance Datum, and one may observe as a corollary of this the existence of considerable areas of almost level ground at about 300 m. It can also be seen that there are indications of a further flattening in the landscape at 240 m (750–800 ft). These surfaces can be traced elsewhere on Bodmin Moor and more generally in South-West England, and if O.S. 1:63 360, Sheet 186, is available, lower surfaces can be located at 170–205 m (550–675 ft) on the S flank of Bodmin Moor, and 90–120 m (300–400 ft) along the coast S of Tintagel. The student must realize, of course, that the positive identification of such features rests upon very detailed map analysis and field study, and that methods such as the projected profile are designed to illustrate, rather than prove, the existence of erosion surfaces.

A falling sea-level implies rejuvenation of rivers. The effects of this rejuvenation are first felt at the coastline and the knick point(s) move progressively upstream. Headward migration of knick points is checked when a highly resistant rock such as granite is encountered. It is therefore to be expected that the deeply incised rejuvenated valley will be found on the edges of the granite outcrop—as at the Devil's Jump (square 1079) while the rivers in the centre of the Moor flow in broad valleys as yet little affected by rejuvenation (*see Further Work—Practical*).

The landscape of Bodmin Moor is thus seen to be a result of (a) the characteristics of the bedrock, (b) the geomorphological processes which have acted upon it, and (c) the stage which these processes have at present reached. To this, the geographer would wish to add that the landscape of today owes much to human activity over the past 3000 years.

Fig 174 Rough Tor from the NW

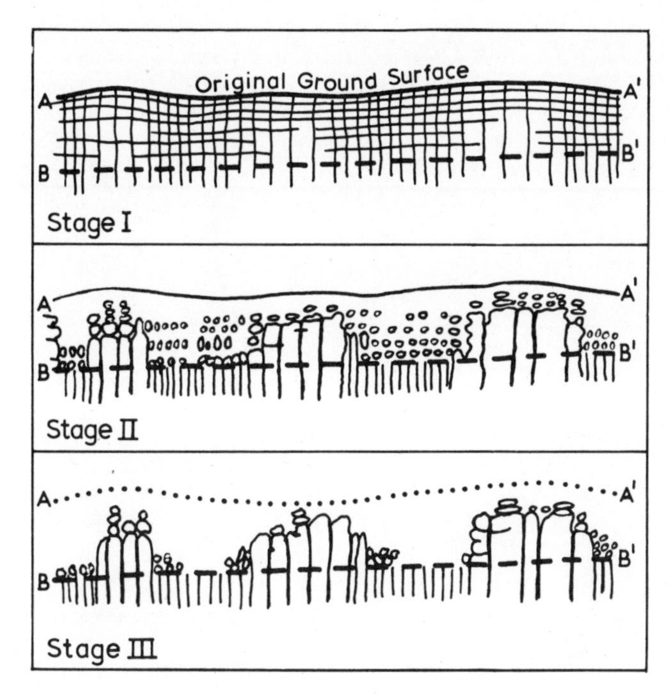

Fig 175 Stages in evolution of tors (after Linton)

FURTHER WORK

Practical
(from O.S. 1:63 360, Sheet 186, Bodmin and Launceston)

(1) Draw long profiles and appropriate cross-sections of rivers draining S from Bodmin Moor. Can you locate the presence of knick points, and breaks of slope on the valley sides? How can you reconcile the facts thus presented with the features of a drowned coastline presented by the mouth of the Fowey River (*Study B8*) into which these streams empty.

(2) Compare the landscape features of Bodmin Moor with the Kingairloch area and area S of Ballachulish on the W shores of Loch Linnhe (O.S. 1:63 360, Sheet 46). What similarities and differences do you observe between these areas which have granite as a bedrock? Can you explain the differences?

Reading

*W. G. V. Balchin: *Cornwall*, British Landscape through Maps Series, Geog. Ass.
Hardy & Monkhouse: *Physical Landscape in Pictures*, pp. 2–3
*L. A. Harvey & D. St. Leger-Gordon: *Dartmoor*, Collins, 1963
Holmes: *Principles of Physical Geology*, *Ch. V, pp. 102–107; Ch. XIV, pp. 615–617
D. L. Linton: 'The Problem of Tors', *Geog. Journal*, vol. 121, 1955
Monkhouse: *Landscape from the Air*, p. 14
Scovel: *Atlas of Landforms*, pp. 52–55
*Sparks: *Geomorphology*, Ch. 7, pp. 149–154; Ch. 3, pp. 33–34
*Strahler: *Physical Geography*, Ch. 31, pp. 484–488
Trueman: *Geology and Scenery*, Ch. 21, pp. 285–293

The Ashover Valley, Derbyshire — Study E6

EROSION AND DRAINAGE PATTERNS IN FOLDED STRATA

O.S. 1:25 000 (4 cm to 1 km *or* 2½ in to 1 mile)
Sheet SK 36

EXERCISE

(1) Draw a contour section from 330610 on the map extract (*Fig 176*) to the Triangulation Point at 359636.

(2) Use the map extract, the field sketch (*Fig 177*) and your contour section in preparing a description of the relief features of the valley.

Your description of the valley should include reference to the following points:

(a) the orientation of the valley, i.e. the direction in which the R. Amber flows across the map extract;

(b) the amplitude of relief: from 300 m (1000 ft) at the crest of the SW side of the valley to the river level of about 150 m (500 ft);

(c) the steep slopes (up to 25°) on each side of the valley between 210 m and 275 m (700 ft and 900 ft contours), culminating in places in sheer outcrops of bare rock;

(d) the gentler slope between 180 m and 215 m (600 ft and 700 ft contours) on the NE side of the valley, with the R. Amber itself flowing in a relatively steep-sided inner valley;

(e) the existence of a ridge within the valley, parallel to the river on its SW side, running from 354619 to 341631.

There are clearly features in this valley which suggest that in explaining its evolution one must look further than the straightforward erosion of a valley by the river, and this is confirmed by field observation. From the view

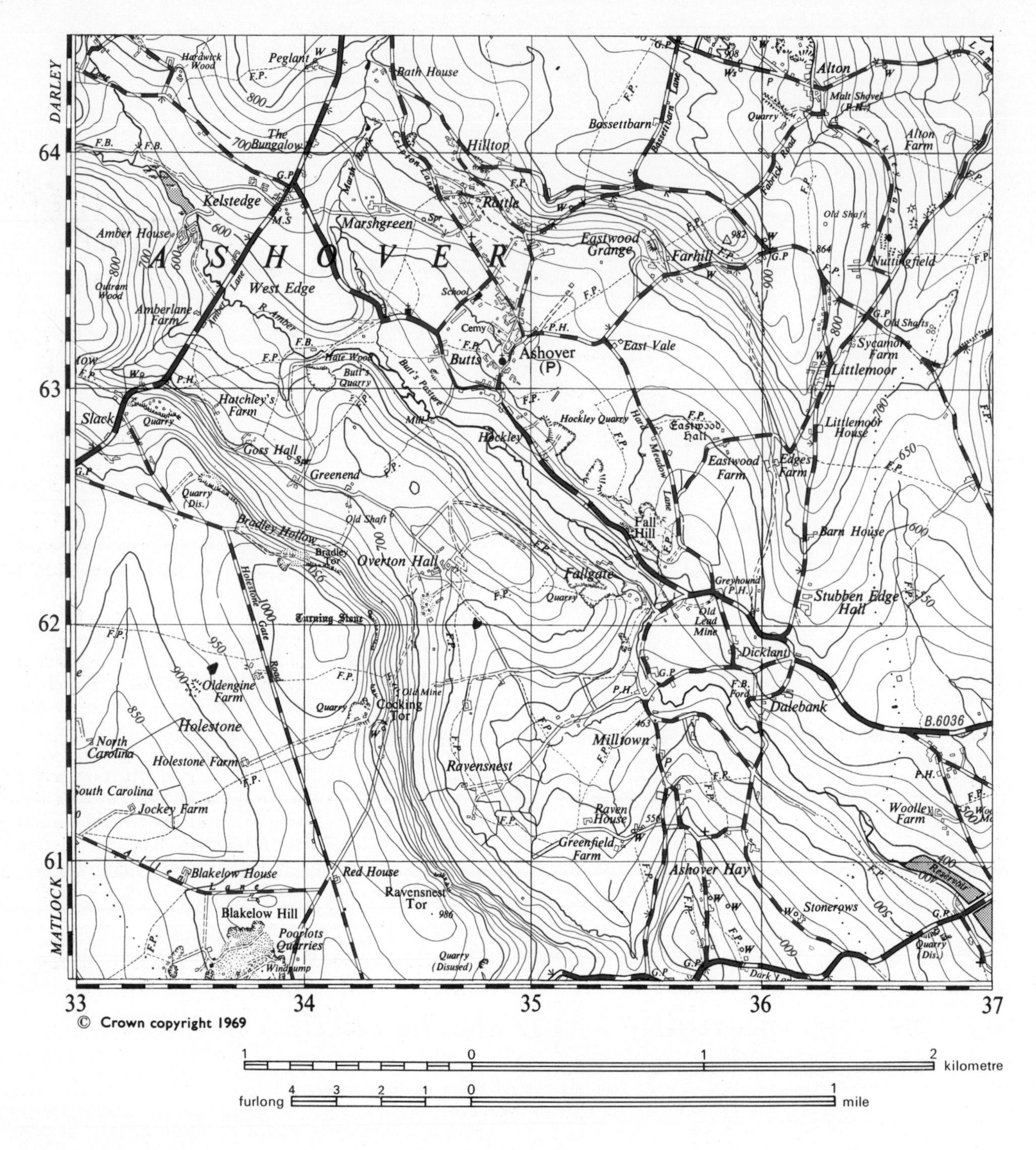

Fig 176 O.S. Ashover
Scale 1:25 000 (4 cm to 1 km *or* 2½ in to 1 mile) Contour interval 25 ft

shown in *Fig 177* one can see, in the field, that the rocks which outcrop on the southern side of the valley appear to be almost black and divided by horizontal and vertical joints into huge quasi-rectangular blocks. In fact, when this rock is broken with a hammer it is seen that the black colouration is only a superficial product of exposure and that the rock itself is a yellowish, coarse-grained sandstone or grit. By contrast, the rock exposed in Butts quarry at the NW end of the central ridge (341630) is greyish white in colour, and closer examination shows this to be a form of limestone. There is thus clear evidence that more than one rock type occurs within the valley, so one turns to the geological map for further information (*Fig 178*).

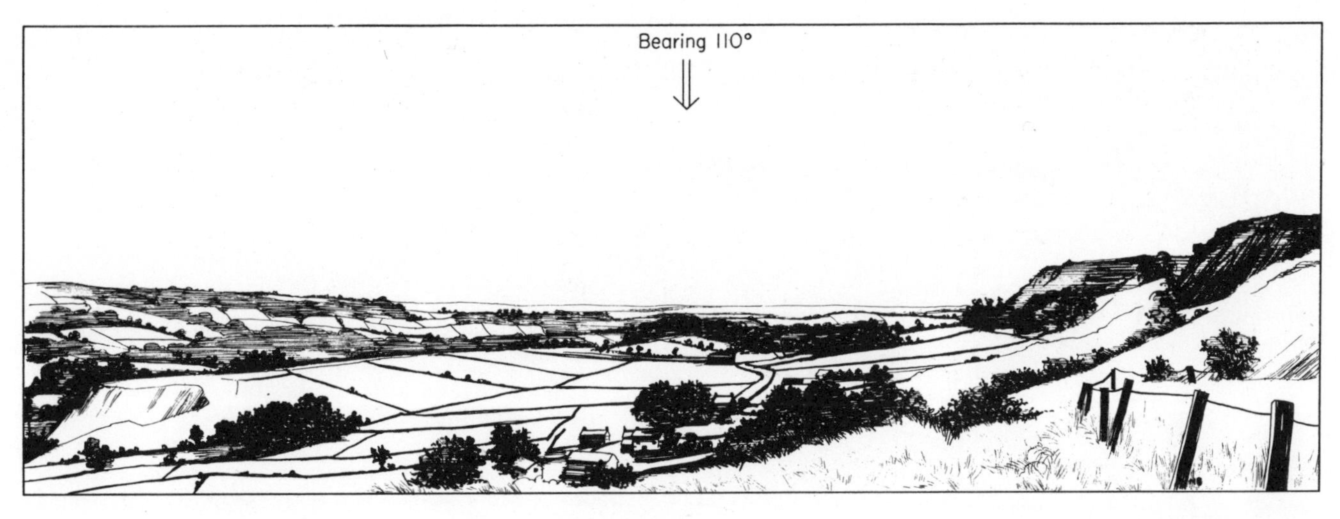

Fig 177 Field sketch of Ashover valley (View point 335628)

The Ashover Valley is situated on the E flank of the Pennine anticline within the area where the main surface rocks belong to the Millstone Grit Series, with a regional dip towards the E, underlain by shales and Carboniferous Limestone. The geological map of the Ashover area indicates that, in this valley, the shales and Carboniferous Limestone are exposed at the surface, and reference to the arrows indicating the dip of the strata shows that this is a minor structural dome with the rocks dipping outwards in all directions from a point near 348627.

(3) Mark the junctions of strata on your section (*Ex. 1*) and use the indications of dip to suggest the relationship of rock strata to relief. *Note*: The Tuff shown in the centre of the valley is not a stratified sedimentary rock, but a compacted volcanic ash of Carboniferous age. The alluvium of the river valley is only a very thin superficial deposit.

It should now be clear that the steep slopes noted between 210 m and 275 m (700 ft and 900 ft contours) are the result of the outcrop on the valley sides of the resistant Ashover Grit forming inward facing scarps; the gentle slope between 180 m and 215 m (600 ft and 700 ft contours) in the NE of the valley coincides with the exposure of Butts Shale; and the central ridge is to be explained by the outcrop of Carboniferous Limestone. The R. Amber follows the central line of uplift cutting part of its valley in the Tuff. In the Ashover Valley, therefore, we have an example of a landscape produced by the erosion of upfolded strata.

The most significant point of general application to emerge from this study is that a zone of structural uplift has been so modified by erosion as to produce a valley. This is sometimes referred to as **inverted relief**, and in folded regions it is more common to find low ground occurring at the crest of anticlines than to find examples of uneroded anticlinal ridges. As further examples of this inversion of relief one may cite:

(a) The Vale of Pewsey, W of Devizes, Wiltshire (*Fig 179*)
(b) The Vale of Wardour, SW of Wilton, Wiltshire
} O.S. 1:63 360 Sheet 167

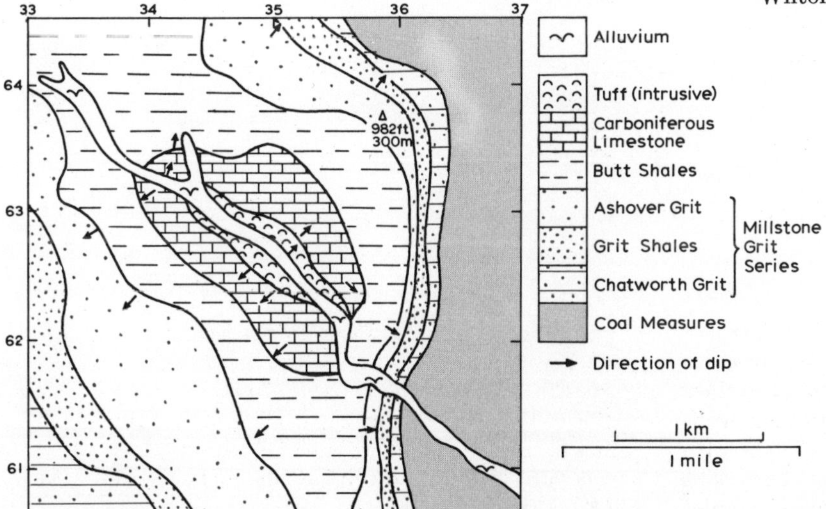

Fig 178 Ashover Valley geology

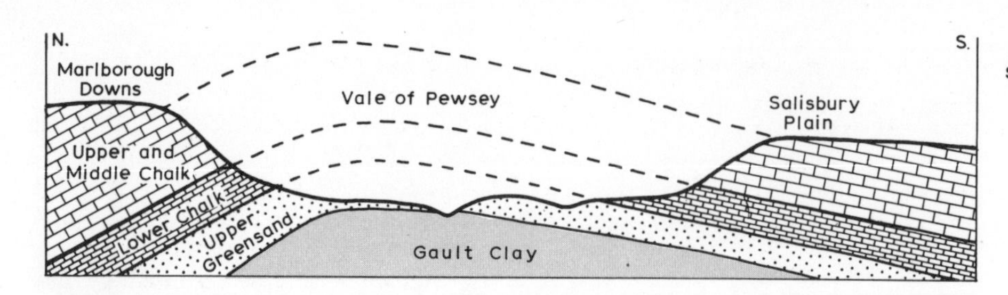

Fig 179 Generalized section through eroded anticline of Vale of Pewsey, Wiltshire

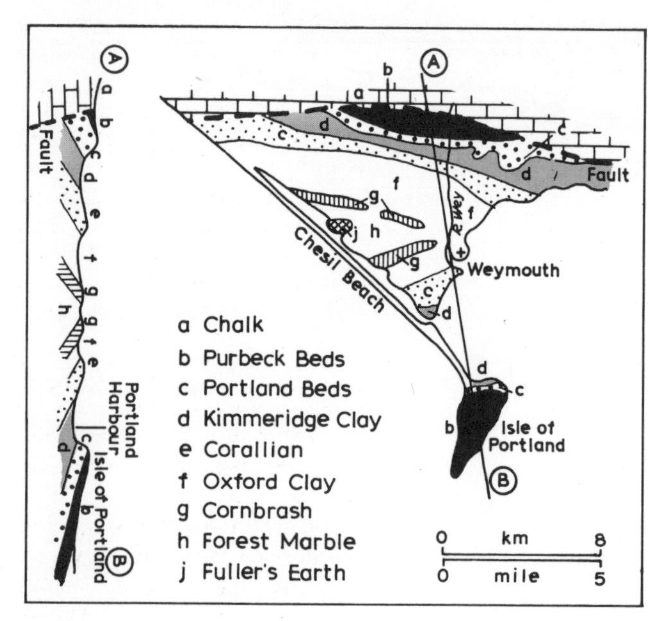

Fig 180 Geological sketch map and section through Weyland, Dorset

(c) Weyland—the district extending from the S edge of the Chalk in Dorset to the Isle of Portland (*Fig 180*). — O.S. 1:63 360 Sheet 178

On a larger scale, the Weald of Kent and Sussex shows similar features and study of maps, books and photographs will reveal many further examples both in Britain and overseas. The maps (*Fig 179* and *180*) show that each example has its own characteristic relationship between structure and relief; the common element is that the greatest degree of erosion has occurred along the anticlinal crest. The simplest explanation for such a phenomenon is that the process of folding must entail some extension, and thus weakening, of strata in the anticlinal crest. Conversely there will be compacting and therefore strengthening in the synclinal troughs. Weathering and erosion are able to make their greatest impression in the crest area with the results that have been cited. The relationship between folded structure and relief can be clearly seen in the photograph of an eroded anticline in Iran. (*Fig 181*). One characteristic which may be observed in this photograph, as well as in

Fig 181 Eroded anticline, Iran

most of the British areas cited above, is that the ridges or inward facing scarps marking the flanks of the anticlines tend to approach each other as the anticline **pitches**. The degree of folding dies out along the anticlinal axis, and as it does so the exposure of older rocks in the centre of the upfold becomes narrower and the flanking slopes unite (*Fig 182*).

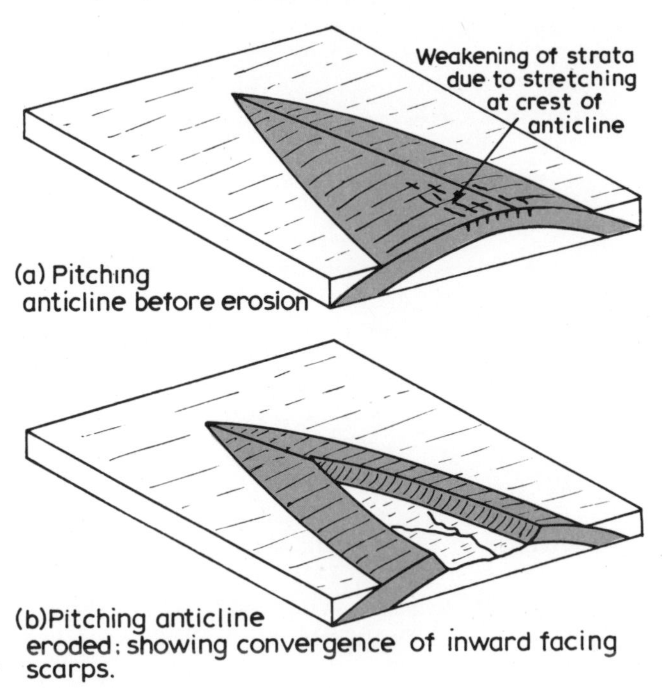

Fig 182 Erosion of a pitching anticline

The relationship between drainage patterns and folded structures is often a complex one. It is, of course, unthinkable that a stream should run along the crest of a newly folded anticlinal ridge. So where we find present-day streams following the anticlinal axis, as at Ashover and in the Vale of Pewsey, two alternatives are possible. Either the stream has developed by headward erosion during the present phase of denudation, or the stream originated on a surface—either erosional or depositional in origin—well above the present valley level. In the latter case the landscape is **polycyclic.** And since no stream could have started its life by flowing up one side of a fold and down the other side, the same argument must apply to a river like the Wey, N of Weymouth, which flows at right angles across the anticlinal axis.

As outlined above (*p. 42*) a stream which can be shown to have its origin directly from the inclination of the rocks is termed **Consequent,** e.g. a stream flowing down a dip slope. Tributaries to such streams which develop along lines of weakness such as exposures of less resistant rocks are known as **Subsequent.** This terminology has been further extended by the use of:

Obsequent: streams flowing in the opposite direction to the consequent river, as in the case of streams flowing away from the foot of a scarp;

Extended consequent: applied when a consequent river is lengthened by flowing across a newly elevated and exposed coastal plain;

Secondary consequent or **Resequent**: streams flowing down the dip slope, parallel to, but of later origin than, consequents;

Insequents: rivers which have no systematic orientation, and whose courses are unrelated to the bedrock.

Geomorphology research has shown, however, that many areas have a complex erosional history, so that it is only after prolonged research at an advanced level that the terms consequent, subsequent, etc. can safely be applied to portions of a river system.

In regions of folded or tilted strata it is normal to find that there are two dominant lines of stream development: (i) parallel to the dip of the rocks, and (ii) parallel to the strike. Since these directions are at right angles, the rectilinear drainage pattern which results, though with many minor irregularities, is known as **trellis drainage.**

FURTHER WORK

Practical

(1) Paint vertical lines on the edge of a piece of plastic foam or sponge. When this is bent to represent folding, the lines will clearly demonstrate the extension of joints at the crest of an anticline.

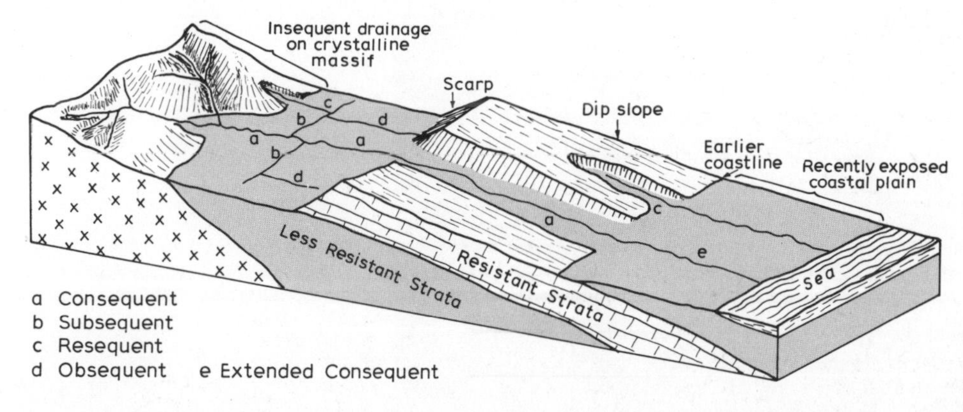

Fig 183 Schematic diagram to show terminology of rivers

(2) Make two simple models to show (a) an anticline when first formed, (b) the same area after erosion has produced 'inverted relief'. *Note*: Plastic foam sheets of differing colours, suitably mounted, form a good basis for such a model.

(3) Study maps of the Appalachian Ridges and Valleys, e.g. those in *Atlas of Landforms*, and draw annotated maps to show their significant physical features.

Reading

Carter: *Land Forms and Life*, Section 15, pp. 119–125; Section 16, pp. 126–135

Dury: *The Face of the Earth*, Ch. 3, pp. 19–30

Gresswell: *Geology for Geographers*, Ch. 4, pp. 54–65

——: *Physical Geography*, Ch. 17, pp. 198–204

Holmes: *Principles of Physical Geology*, Ch. 19, pp. 566–569; *Folding, Ch. IX, pp. 206–214

Monkhouse: *Principles of Physical Geography*, Ch. 6, pp. 144–148

——: *Landscape from the Air*, p. 5

Scovel: *Atlas of Landforms*, pp. 32–41

*Sparks: *Geomorphology*, Ch. 6, pp. 101–111

*Strahler: *Physical Geography*, Ch. 29, pp. 448–465; Ch. 30, pp. 471–474, Ex. 1, p. 482

Trueman: *Geology and Scenery*, Ch. 7, pp. 91–101

Wooldridge & Morgan: *An Outline of Geomorphology*, Ch. XIV, pp. 180–188

The Crater of Kilauea, Hawaii

Study E7

VOLCANIC ACTIVITY

U.S. Geological Survey 1:24 000 (4·2 cm to 1 km *or* about 2½ in to 1 mile)

Volcanoes have always been regarded with awe by man, and have often been invested with superstitious significance. Belief in a fiery hell beneath the earth must surely owe its origin to fearful observation of craters in eruption. Modern man still has good reason to fear the virtually unpredictable and often destructive release of energy from an erupting volcano—energy which may take the form of white hot, flowing lava; catastrophic explosion; or belching poisonous gases, as well as many other phenomena.

The crater of Kilauea, illustrated on the map and photograph (*Figs 184* and *185*) lies on the SE slope of the larger volcanic dome of Mauna Loa on the island of Hawaii. Mauna Loa rises to an altitude of 4170 m (13 680 ft) above sea-level which alone marks this out as a major volcanic peak. Yet only the top half of this great structure rises above sea-level. The ocean floor in mid-

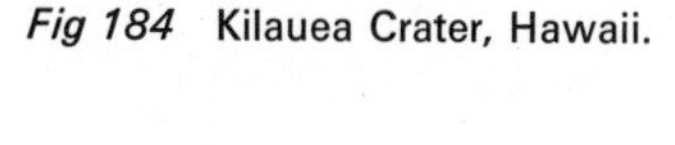

Fig 184 Kilauea Crater, Hawaii.

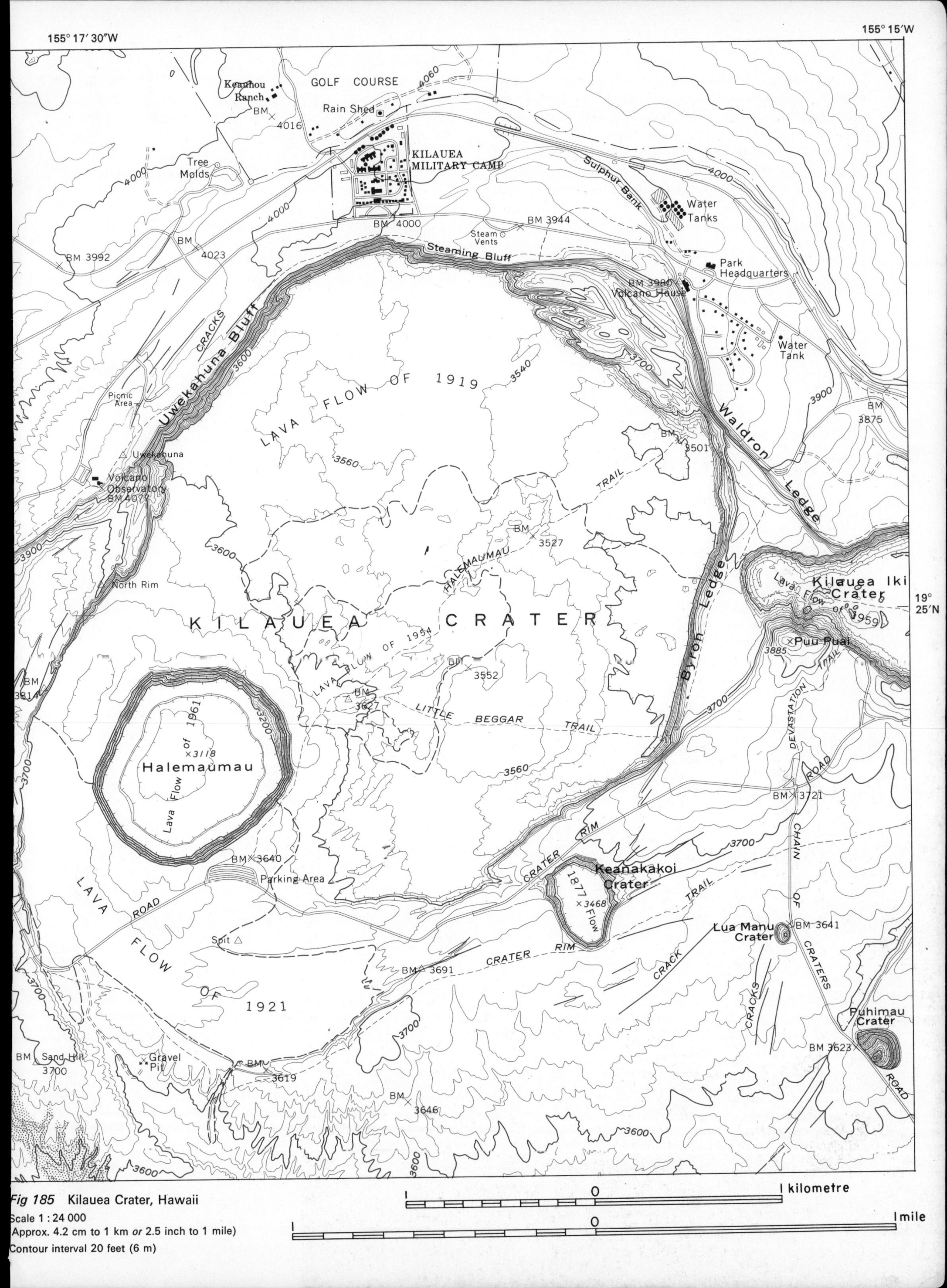

Fig 185 Kilauea Crater, Hawaii

Scale 1 : 24 000

(Approx. 4.2 cm to 1 km *or* 2.5 inch to 1 mile)

Contour interval 20 feet (6 m)

Pacific is some 5180 m below the surface, thus the total height of Mauna Loa is over 9350 m (30 000 ft)—a colossal accumulation of volcanic material which probably began in mid-Tertiary times, 25 million years ago.

The average slope of the upper part of the dome of Mauna Loa (seen in the background of *Fig 184*) is 1 in $6\frac{1}{2}$. What angle of slope is this? Can this be checked from the photograph? How do the lower slopes of Mauna Loa differ from the upper slopes?

EXERCISE

Draw a contour section across the map from SW to NE passing through the depression of Halemaumau. Relate your section to the photograph, and either draw a sketch of the photograph or place tracing paper over the photograph and show the location of the following features marked on the map:

Halemaumau depression; the lava flows of 1921 and 1954; the Crater Rim Road; Keanakakoi Crater; Waldron Ledge; Steaming Bluff; Gravel Pit.

This volcano occurs in a region of high rainfall, as is suggested by the well-developed gulleying on the outer slope of Kilauea. Why does there appear to be little vegetation in spite of this high rainfall?

The depression, or pit, of Halemaumau often contains a lake of bubbling, molten lava. The pit has varied in size from 130 m–400 m ($\frac{1}{4}$–$\frac{3}{4}$ mile) in diameter. At times the surface of its lava lake has been 400 m (1300 ft) below the rim—at other periods the lava rises and spills over on to the main crater floor as in 1921 and 1954. In the former eruption a stream of molten lava overflowed from the main crater forming a thin 'river' now solidified which can be seen in the foreground of the photograph. Lava, gases and steam may also issue from the smaller vents within the crater rim and from small subsidiary craters on the flanks such as Keenakakoi. The photograph suggests that lava erupting from this volcano flows in quite a liquid state, spreading out into a fairly smooth sheet before cooling and solidifying, and such is in fact the case. Other volcanoes erupt in different ways, producing different landforms which are considered below.

Whence comes the energy for volcanic activity? It has been shown above (*p. 93*) that the temperature of the earth increases towards the centre and that at the base of the crust temperatures may be high enough to melt rocks if they were at atmospheric pressure. Part of this heat, perhaps about half, can be attributed to the heat retained from the time when the whole earth was a molten mass. The remaining heat is the result of radio-activity within the earth's rocks and is being constantly generated. Here then is the source of energy, but why should it be released only at certain parts of the earth's crust?

Fig 186 World distribution of active volcanoes

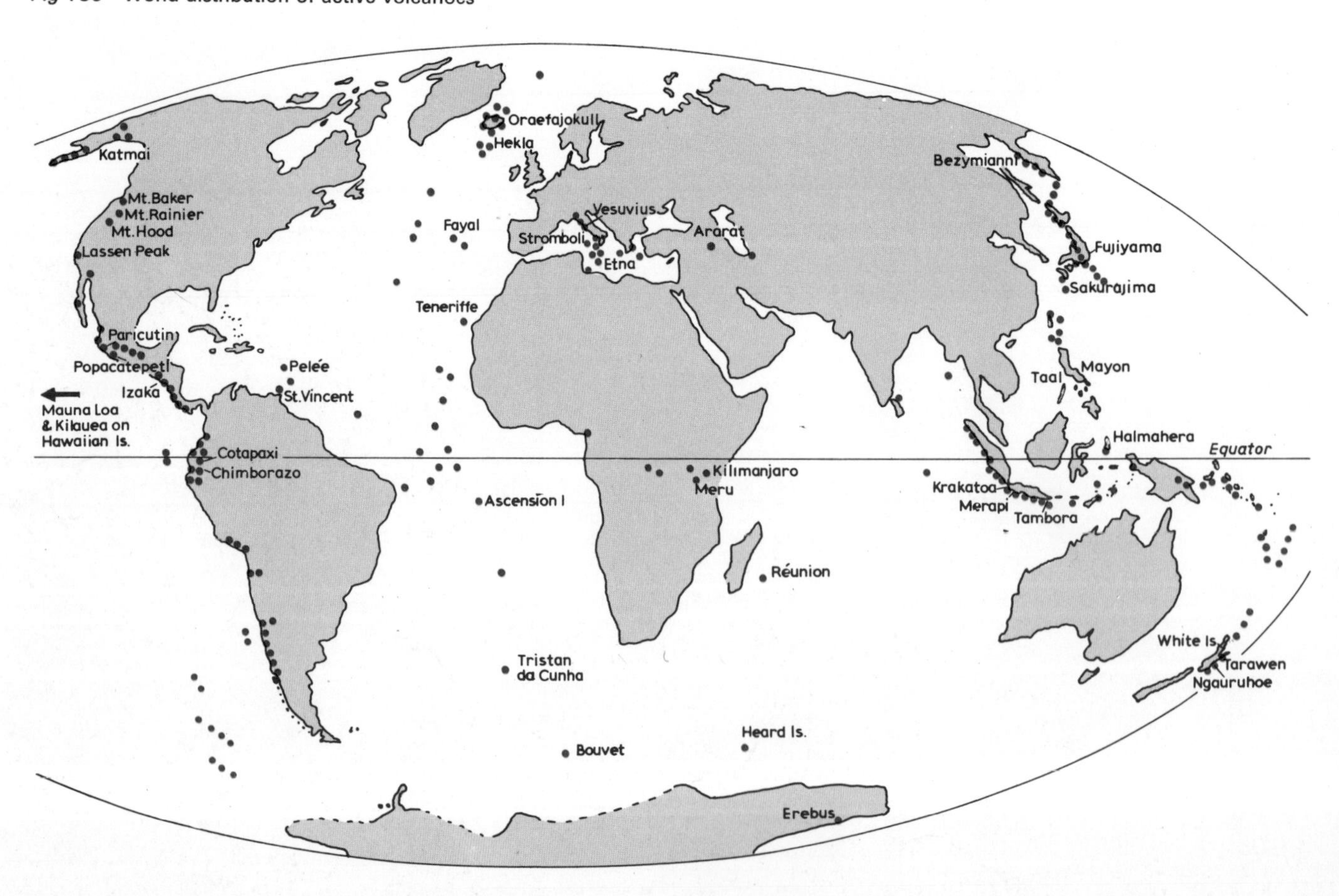

If the world map showing the distribution of recently active volcanoes (*Fig 186*) is studied, it will be seen that these occur first around the margins and in the centre of the Pacific Ocean; secondly in an arc extending from New Guinea through Java and Sumatra to the Andaman Islands; thirdly in association with the great rift valleys of Africa; fourthly in an E–W belt through the Mediterranean, and finally along the mid-Atlantic Ridge from Iceland to Bouvet Island and in the Eastern Caribbean. These are all regions of recent mountain building activity and tectonic instability where deep cracks are likely to occur in the rocks of the earth's crust. If a crack or zone of weakness were to extend to a depth of 60 km, the pressure at such a depth would be 17 000 times that of the surface. A local release in pressure allows the intensely hot rocks to become molten and they will be forced up the crack—such liquified rock is described by the general term of **magma**. As the magma rises it begins to cool and may solidify before reaching the surface, spreading out into a deep-seated mass and cooling slowly to solidify with visible crystals as a **batholith**, perhaps of granite. Alternatively the magma may solidify in narrow sheets parallel to or cutting across the bedding planes of the surrounding rocks to form a **sill** or **dyke** (*Figs 187* and *188*). In this case the cooling is often more rapid and the crystals formed are very small. However, the temperature of the magma is such that in certain places it will remain sufficiently liquid to reach the surface, and as the pressure decreases with ascent, gases bubble out from the magma increasing its mobility.

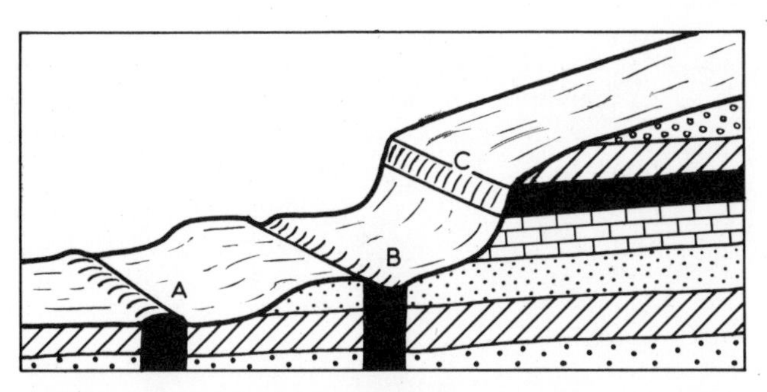

Fig 187 Dykes (A and B) and Sill (C)
Note that dyke A stands out from less resistant rock, Dyke B is surrounded at the surface by more resistant rock and is consequently slightly below the general level

But not all magma has the same properties. The Hawaiian volcanoes of which Kilauea is an example are characterized by a very free-flowing lava. In part this fluidity is a consequence of the presence of large volumes of volcanic gases, particularly steam, which by their physical presence and chemical reactions maintain the lava at the high temperatures (*c.* 900–1200°C) necessary for liquid flow. The chemical composition of the lava has an important effect upon its ability to flow easily. Kilauea lava, like all those of the Hawaiian volcanoes has a relatively low proportion of free quartz (SiO_2) and is

Fig 188 Win Sill, Northumberland, surmounted by Hadrian's Wall

chemically basic in character. Such basaltic lavas always tend to spread far outward from the central vent or fissure, and cool and solidify with a low angle of slope. Lavas of this type emerging from a series of long fissures have spread to cover at least 260 000 km² of the states of Washington and Oregon in the NW of the U.S.A. and similar lava flows are found, for example, on the Deccan Plateau of India, in Antrim (Northern Ireland) and the Inner Hebrides (*Fig 189*).

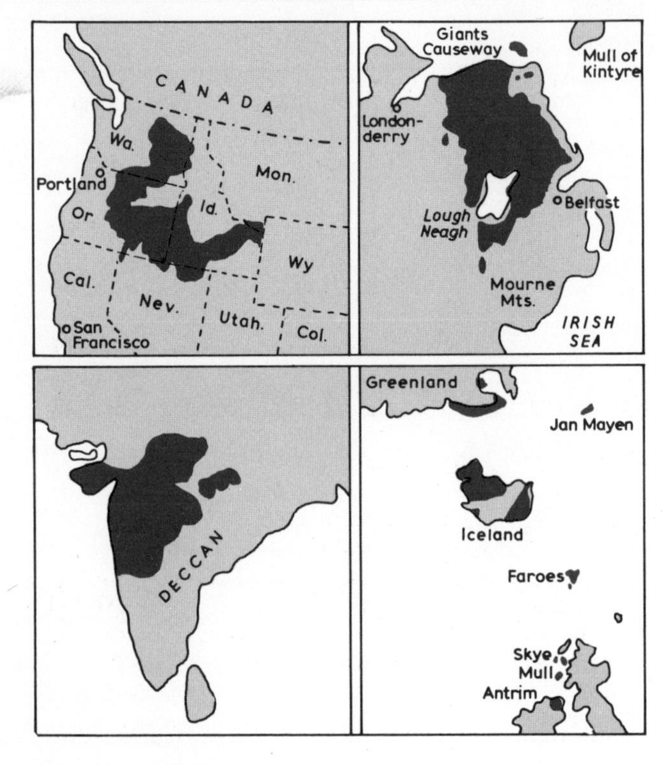

Fig 189 Basalt Plateaus

Where silica rich, acidic lavas erupt at the surface a very different type of eruption and land form results. The characteristic of such lavas is to solidify quickly so that there is a tendency for the vent of the volcano to become blocked or plugged by hardened lava. Pressure builds up beneath such a plug, hot gases melt the plug from the base and eventually the plug is destroyed or a new outlet formed, often with explosive force. Great clouds of pulverized ash are thrown high into the air to settle on the land for miles around, and thick or viscous lava oozes from the vent and spreads slowly outward. Volcanic cones resulting entirely from the extrusion of acidic lava are far steeper than the basic volcanic cones of the Hawaiian Islands since the acidic lava flows much more slowly and solidifies more quickly. Extinct volcanoes of the Auvergne district of Central France (*Fig 190*) are of this type and exhibit characteristic convex forms still visible after long periods of denudation.

The explosive activity associated with most volcanic eruptions is largely due to the releases of steam and gases from the upwelling lava. Explosions hurl both molten lava fragments and fragmented ash high into the air, and such material (**pyroclasts**) may be thrown upwards and outwards with sufficient force to fall to the surface some kilometres from the vent, and can accumulate into deposits of considerable thickness, the fragments of which eventually become cemented together to form **tuff**. The most often depicted volcano is of the type built up by successive deposits of ash and lava to form a **composite cone**—examples of which include Vesuvius and Fujiyama (*Fig 191*).

The possible variations in volcanic eruptions and in the resulting land forms is almost infinite and this study only considers a few types and outlines some general principles. *Study E8* exemplifies the process of caldera formation in which much of a visible cone collapses into a 'cauldron' of volcanic material below.

Fig 190 Plateau des Domes from SW, Puy-de-Dome, France

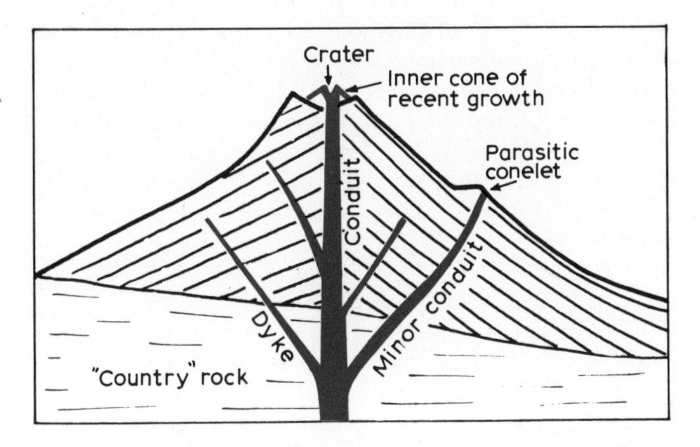

Fig 191 Composite Volcanic Cone, formed of layers of ash and solidified lava

FURTHER STUDY

Practical

(1) Study as many photographs of volcanoes as possible. Draw sketches based on the photographs and use annotation to indicate significant features.

(2) If at all possible try to see Haroun Tazief's remarkable film *Volcano*. How many different kinds of volcanic activity are shown in this film?

(3) Most of this Study has been concerned with recently active volcanoes. What will be the effect upon volcanic features of long continued periods of denudation? Remember that solidified lavas will normally be relatively resistant to erosion, and certainly more resistant than consolidated ash. Relate your conclusions to maps and/or photographs of Arthur's Seat, Edinburgh; the town of Le Puys in Central France; Ship Rock, New Mexico, etc.

Reading

Anderson: *Splendour of Earth*, pp. 158–175
Carter: *Land Forms and Life*, Section 19, pp. 155–159
Dury: *The Face of the Earth*, Ch. 5, pp. 46–55
Gresswell: *Geology for Geographers*, Ch. 2, pp. 14–38
——: *Physical Geography*, Ch. 32, pp. 429–444
Hardy & Monkhouse: *Physical Landscape in Pictures*, pp. 20–27
*Holmes: *Principles of Physical Geology*, Ch. XII, pp. 287–344
Horrocks: *Physical Geography and Climatology*, Ch. 2, pp. 21–30
Monkhouse: *Principles of Physical Geography*, Ch. 3, pp. 46–72
——: *Landscape from the Air*, p. 8
Scovel: *Atlas of Landforms*, pp. 62–74
*Sparks: *Geomorphology*, Ch. 7, pp. 149–154
Strahler: *Physical Geography*, Ch. 31, pp. 484–498
H. Tazieff: *Volcanoes*, Prentice-Hall, 1962
G. Walker: 'Mt. Etna', *Geog. Mag.*, vol. 40, No. 11, March 1968, and other similar descriptive articles.

Crater Lake, Oregon, U.S.A.

TYPES OF LAKES

Study E8

U.S. Geological Survey 1:62 500 (1·6 cm to 1 km *or* 1 in to 1 mile)
Crater Lake National Park

EXERCISE (*Fig 192*)

(1) What is the maximum diameter of this lake? What is the surface area of Crater Lake? This can be approximately measured by drawing squares with sides 0.5 km long in scale distance on a piece of tracing paper. Each square represents 0.25 km^2. Place tracing paper over map and mark whole squares covered by water. Then mark those squares more than half covered by water; ignore squares with less than half water. Add up the number of squares marked—and multiply by 0.25 km^2 to obtain approximate area of lake (*Fig 193*).

Note: on an O.S. map the kilometre grid squares may be conveniently sub-divided.

Since this lake is almost circular, how else could you obtain its area?

(2) Draw a cross-section from E to W across the map through The Watchman, Wizard Island and Mt. Scott. Indicate both the water surface and the lake floor as shown by the depth contours indicated.

(3) Study Kerr Valley (SE of the lake). What indications can you see that this is a glacial feature? Where is the corrie in which the ice collected before moving downhill to carve this valley.

(4) Red Cone, Grouse Hill, and Mt. Scott are hills rising from the slopes around Crater Lake. Bearing in mind the words 'Crater' and 'Cone', what origin would you suggest for these hills?

The occurrence of names like Crater Lake, Red Cone, Pumice Point (N of Lake) and the general forms of land

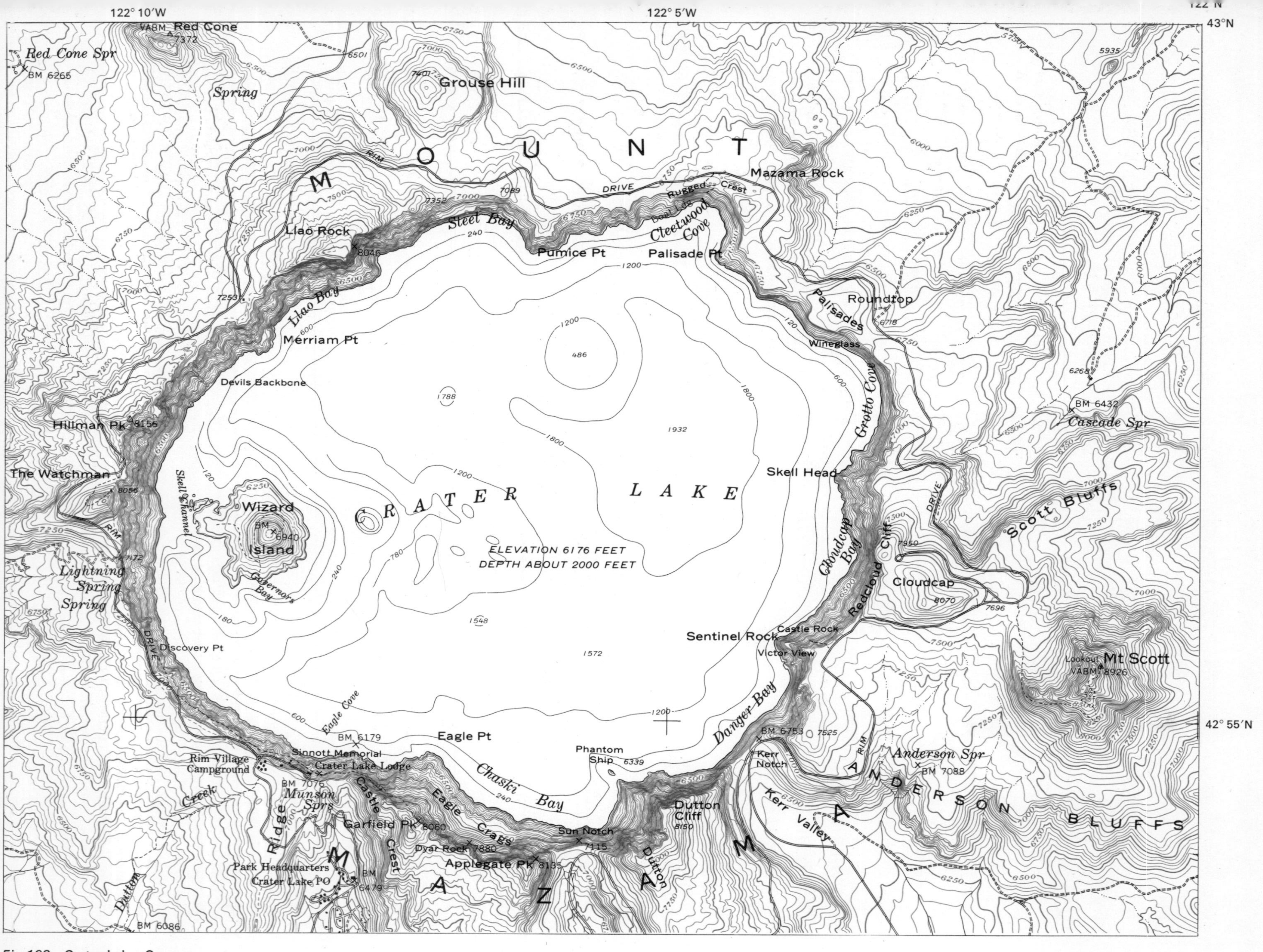

Fig 192 Crater Lake, Oregon
Scale 1:62 500 (1·7 cm to 1 km or 1 in to 1 mile) Contour interval 50 ft

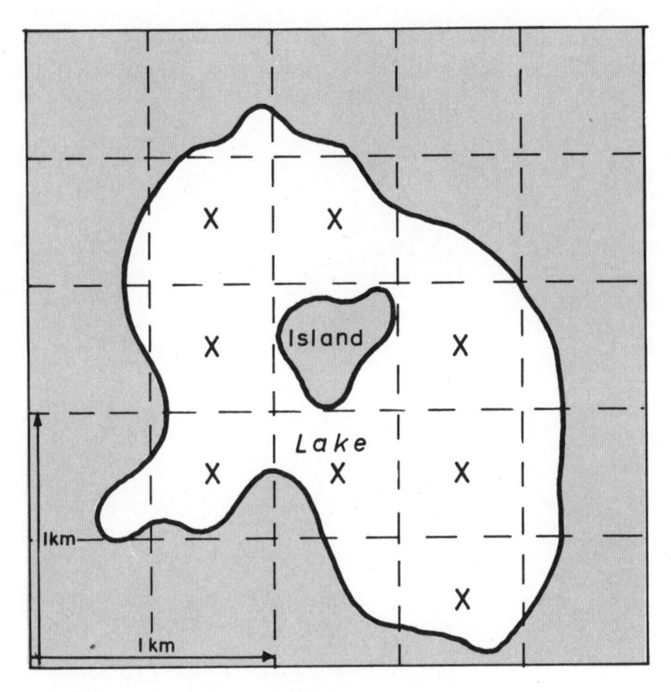

Fig 193 Each square = 0·25 km². Squares marked × are more than half water.
8 squares are so marked. $8 \times 0{\cdot}25\ \text{km}^2 = 2\ \text{km}^2$
(N.B. Greater accuracy is obtained with this method by using smaller squares.)

in this area should leave the student in little doubt that this is a feature of volcanic origin, and the cross-section drawn in *Exercise 2* above may suggest that this area was once a much higher volcanic peak which in some way has lost its top. Such a hypothesis can be confirmed by field observations of the nature and arrangement of beds of lava, pumice, etc. on the existing sides of the lake. The significance of *Exercise 3* above lies in dating the destruction of the peak, since Kerr Valley is clearly only the lower portion of a glaciated valley which must have been supplied with ice from corries at a considerably higher level on a peak which has now disappeared. Hence the destruction of the peak post-dates the glacial epoch (the Pleistocene) during which the valley was carved. Radiocarbon dating of carbonized wood from trees destroyed by the final eruption of this great volcano, shows that this occurred about 8000 years ago, and glacial and geological evidence suggests that a cone at least 3500 m (12 000 ft) high then existed on the site of Crater Lake.

The probable sequence of events was:

(a) Catastrophic eruptions pouring out masses of lava and gaseous material from an underlying magma 'reservoir'.

(b) The extrusion of this material left the cone without support, so that it collapsed into the space below, leaving the present great crater—correctly termed a **caldera.**

(c) Subsequent minor eruptions formed the cone of Wizard Island, and possibly other smaller cones which lie below the lake surface.

Crater Lake is an example of one way in which a hollow may be formed on the earth's surface within which water accumulates to form a lake. Although not the only example of a lake in a caldera or extinct volcanic crater, this is not a common phenomenon. Many lakes of varying origins have been referred to in the preceding pages of this book. The formation of the basin in which the water accumulates may be classified into three broad groups: produced by

(a) erosional processes;
(b) depositional processes,
and (c) structural and volcanic processes.

Among lakes owing their origin primarily to erosional processes may be cited corrie lakes; lakes formed in hollows excavated by glaciers in a valley floor or by ice sheets; hollows produced by underground solution, as in some limestone areas; water filled hollows excavated in desert areas by wind action. It may well be that some of these basins have their capacity further enlarged by depositional processes such as the formation of a terminal moraine on the rock barrier ponding up a ribbon lake.

In the case of lake basins resulting entirely from depositional processes a mass of material forms a natural dam which ponds back the lake water. Such a barrier may take the form of a terminal moraine, or a landslide partially blocking a valley. Marine deposition of sand bars, etc. can isolate an area of water or so disrupt land drainage that shallow lakes are formed such as the Haffs of the Baltic coast near the Polish/U.S.S.R. border (*Fig 194*) or the Etangs of the Rhône Delta (*Study A9*). One case of lake formation in which this process has played a part is that of the Norfolk Broads, though these lakes are now known to be of complex origin in which the formation of peat and its extraction by man, as well as post-glacial changes in sea-level, have all played a part.

Fig 194 Formation of lagoons (Haff) by construction of bay mouth bars (Nehrung)

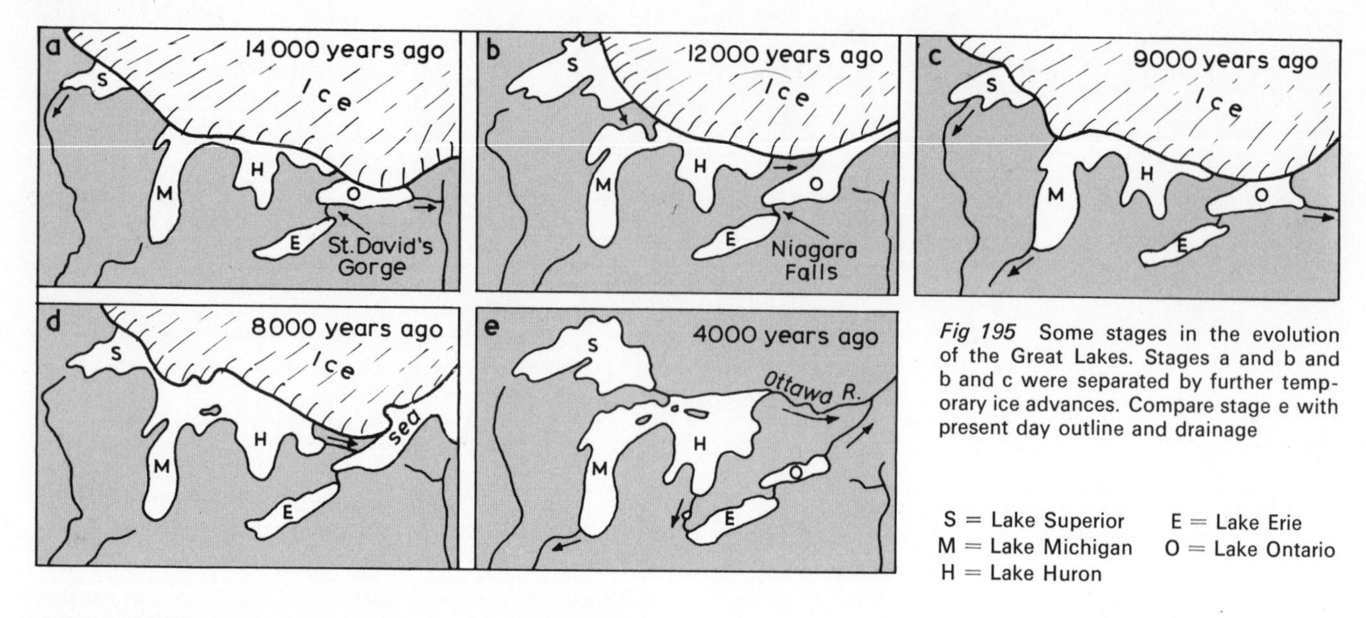

Fig 195 Some stages in the evolution of the Great Lakes. Stages a and b and b and c were separated by further temporary ice advances. Compare stage e with present day outline and drainage

S = Lake Superior
M = Lake Michigan
H = Lake Huron
E = Lake Erie
O = Lake Ontario

Ox bow lakes provide an example in which deposition of river alluvium, as well as the river's power of lateral erosion, combine to isolate a meander loop from the main stream. A mass of ice can form the obstruction damming up a lake as in the ice-dammed marginal lakes cited in *Study C5*. **Kettle hole** lake depressions form where masses of stagnant ice become trapped within boulder clay; as the ice melts the overlying debris collapses into the space so formed to leave a hollow at the surface. In many cases it is difficult, if not impossible, to assign a lake basin solely to one or other of the main categories suggested above. The Great Lakes of North America, for example, have a complex history involving crustal warping, ice sheet erosion, temporary damming by ice, the opening and erosion of various overflows and some degree of morainic blockage.

This section opened with a detailed study of one lake resultant upon vulcanicity. Other crater lakes can be found, and in addition volcanic lava flowing across a valley may form a natural dam. Structural lake basins are widespread, ranging from those like Lough Neagh in Northern Ireland or Lake Victoria in East Africa produced by down warping of the earth's crust, to examples consequent upon faulting such as the Dead Sea, Great Salt Lake or Lake Baikal in Central Asia. In the twentieth century it is tempting to add a further category—that of man-made lakes, for it is in the formation of such vast stretches of water as Lake Mead behind Hoover (Boulder) Dam in Nevada, Lake Nasser behind the Aswan High Dam in the Nile Valley and the conversion of the Zuider Zee to the fresh-water Ijsselmeer, that man's claim to recognition as a geomorphological agent is most clearly seen.

Lakes are essentially temporary features in terms of geological time. Rivers flowing into a lake bring sediment which gradually fills the basin. At the same time the outlet, flowing over the rock or depositional barrier is eroding vertically and thus lowering the lake level, so that eventually the expanse of water is replaced by a plain of lacustrine deposits. It has been estimated for example that, left to itself, Lake Mead will be completely filled with sediment within the next 300 years.

FURTHER WORK

Practical

(1) The summary of ways in which lake basins are formed in the preceding Study is by no means exhaustive and only a few examples are given. Use the references below to compile your own classification of lake basins, giving an example of each type accompanied by an annotated sketch map—many such maps may be constructed with the aid of a good atlas.

(2) Use the 1:63 360 O.S. Tourist map of the Lake District, or the equivalent 7th Series 1:63 360 O.S. maps to work the following exercise which will show stages in the alluvial in-filling of glacial lakes.

(a) Draw a map of the S half of Ullswater, annotated to show areas of alluvial infilling of a deltaic nature at Glenridding (388170), Aira Point (403198) and Sandwick (424198).

(b) Draw a sketch map of Buttermere and Crummock Water annotated to show how alluvial deposition from Sail Beck and Sourmilk Gill has separated a once continuous stretch of water into two lakes. (If preferred, the area between Derwentwater and Bassenthwaite Lake may be considered.)

(c) Draw a map of Upper Borrowdale between Northings 12 and 17. This valley was once a lake.

Reading

Gresswell: *Rivers and Valleys*, Ch. 8, pp. 81–92

Hardy & Monkhouse: *Physical Landscape in Pictures*, pp. 60–63

Holmes: *Principles of Physical Geology*, Ch. XX, pp. 671–675

Monkhouse: *Principles of Physical Geography*, Ch. 7, pp. 154–172

——: *Landscape from the Air*, p. 50

Niagara Falls, Canada–U.S. Border

Study E9

NATURE AND OCCURRENCE OF WATERFALLS

U.S. Geological Survey 1:62 500 (1·6 cm to 1 km *or* 1 in to 1 mile
Niagara River

The Niagara River flows northward for 50 km from Lake Erie to Lake Ontario. In that distance it falls a total of 108 m (from 573 ft above sea-level at Lake Erie to Lake Ontario's surface at 246 ft). This would give an average gradient of approximately 1 in 500 (10 ft per mile) but study of the map extract *Fig 197* shows that such a calculation is meaningless. In the 22 km from Lake Erie to the S limit of the extract the fall is negligible—of the order of 4 m—and similarly at the N edge of the extract the river at Lewiston is only 1 m (spot height 249 ft) above the level of Lake Ontario. Almost all the descent is thus contained within the area of the map extract, and closer study shows that the river falls 13 m (from 560 ft to 520 ft) in the 1.5 km above Niagara Falls; 55 m (160 ft) is accounted for by the falls themselves (517 ft to 357 ft above sea-level), leaving a descent of 35 m (108 ft) between the foot of the falls and Lewiston.

EXERCISE

Draw a long profile to scale along that part of Niagara River shown on the map extract. (Vertical Scale $\frac{1}{10}$ in to 50 ft is used for *Fig 198*. This is a high degree of vertical exaggeration, but is necessary in the present instance.)

Fig 196 Niagara Falls

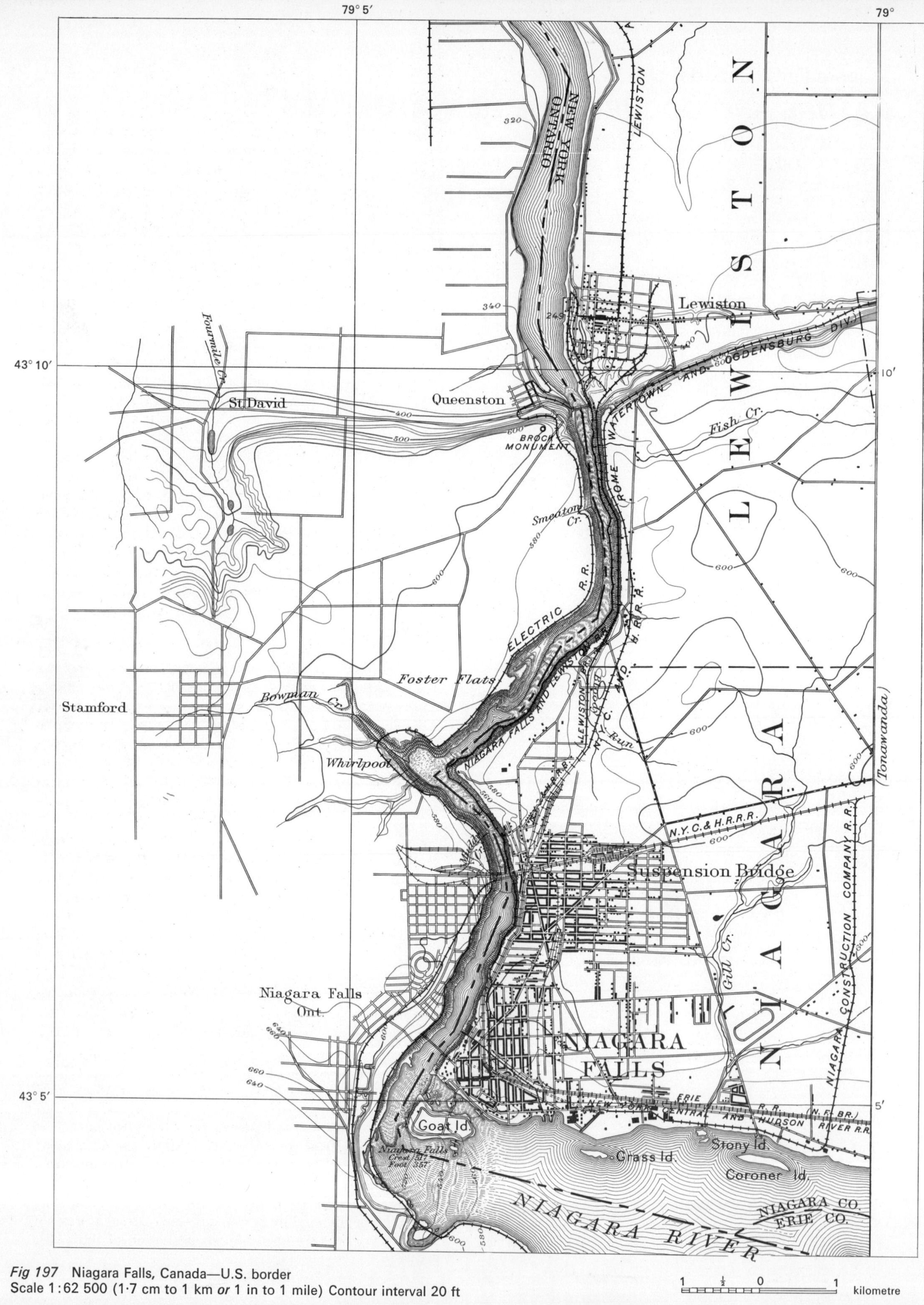

Fig 197 Niagara Falls, Canada—U.S. border
Scale 1:62 500 (1·7 cm to 1 km *or* 1 in to 1 mile) Contour interval 20 ft

1 ½ 0 1 kilometre

1 ½ 0 1 mile

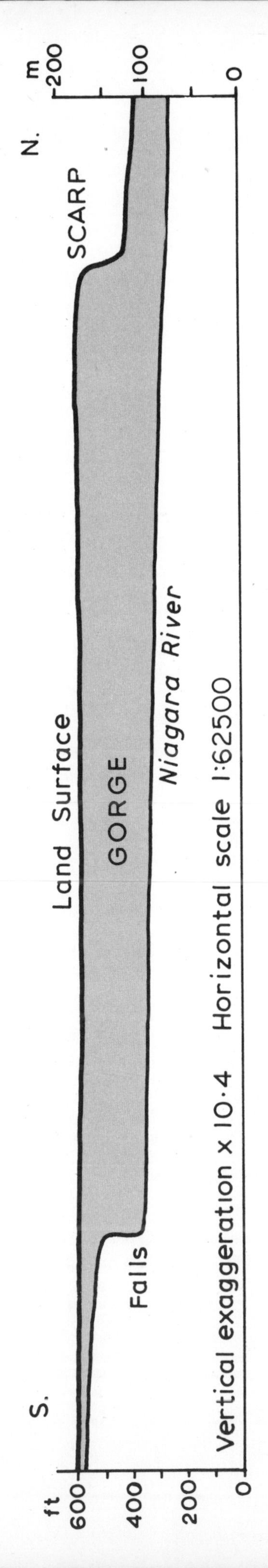

Fig 198 Section along Niagara River and along land surface

If one considers the levels of the land adjoining the river, along its right (E) bank for example, it can be seen that there is a decrease in height from a little below 177 m (580 ft) in the S of the extract to about 107 m (350 ft) to the N of Lewiston. As in the case of the river, this is not an even gradient—in fact almost all the loss in height occurs immediately S of Lewiston where a N facing scarp of some 60 m (200 ft) can be traced running from E to W across the course of the river. The heights of the land bordering the river can be superimposed upon the long profile of the river suggested in the Exercise above (*Fig 198*). Summing up the observations made above it is possible to say that: The Niagara River, 2 km ($1\frac{1}{2}$ miles) wide in the S of the area shown has a very slight gradient until it steepens about a kilometre above the Falls, over which it plunges 49 m (160 ft) into a gorge less than 400 m ($\frac{1}{4}$ mile) wide, some 75 m (250 ft) deep and 11 km (7 miles) long. Within the gorge the river falls 30 m (100 ft) in altitude. The gorge ends at Lewiston where there is a N facing escarpment 60 m (200 ft) in height. *Fig 199* shows some of these points in the form of a block diagram.

One further feature which can be observed from the map demands attention. This is the sharp right-angled bend in the gorge and the feature marked as The Whirlpool. This should be noted in conjunction with (1) the Bowman Creek and its associated gorge which here joins the Niagara River, and also with (2) the embayment into the escarpment south of St. David.

The evolutionary history of the Great Lakes drainage system, of which the Niagara River forms a part, has been reconstructed in considerable details from studies of the land forms, depositional materials and other data (*Fig 195*). For a full account, reference may be made to the list of Further Reading at the end of this chapter. So far as the Niagara River is concerned, its present course was established some 12 000 years ago when water from Lake Erie began to escape northward to Lake Ontario, tumbling over the edge of the escarpment which is formed by the outcrop of a bed of highly resistant dolomitic limestone. The original site of the Falls was thus immediately S of the position of Lewiston, and at that time their height would have been approximately 60 m. By what process have the Falls retreated 11 km (7 miles) in 12 000 years—an average rate of about 1 m (3.08 ft) per year? As the water velocity accelerates over the falls and then rapidly decelerates on reaching the river at the foot, the sudden changes in pressure set up strong erosive action at the base of the falls, thus deepening the channel to produce a plunge pool (*Fig 200*). The same changes in pressure also produce fine spray and back eddies which are very effective in scouring out the less resistant beds beneath the dolomitic 'cap rock'. As this scouring proceeds the cap rock is left without support and blocks collapse so that the position of the falls recedes. Measurements at Niagara in the nineteenth century showed an average recession of 1.2 m per year. Now that the volume of water has been much reduced by the construction of hydro-electric power plants, this recession is slower, but it still causes concern to a town largely dependent on the tourist attraction of the Falls,

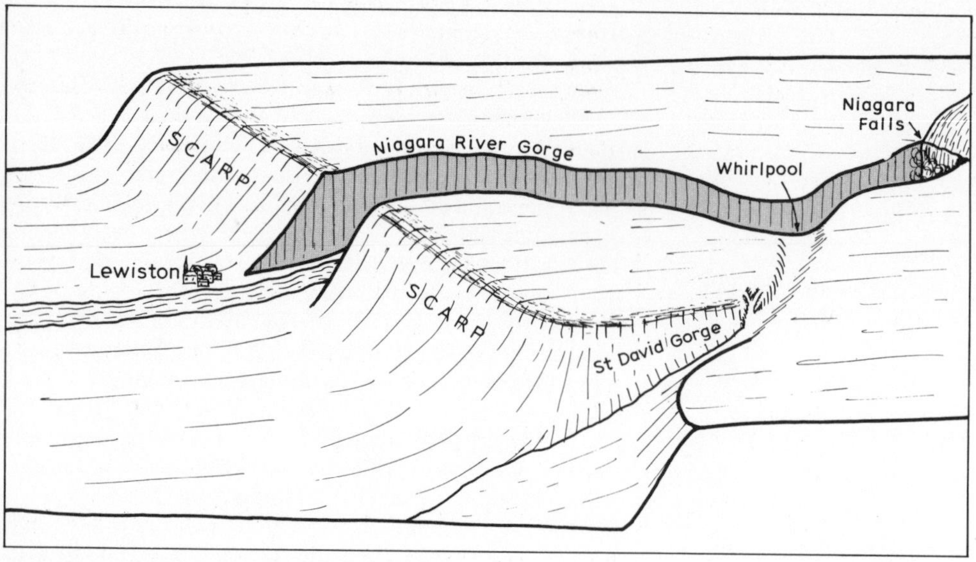

Fig 199 Diagrammatic sketch to show relationship of Niagara Falls and Gorge to St David Gorge

and in 1966 the army was called in to attempt to reinforce blocks of the cap rock which seemed in danger of collapse.

The sharp bend in the gorge at The Whirlpool and the associated features of the Bowman Creek Valley and the St. David re-entrant are attributable to the existence of an earlier course of Niagara River to the W of its present site. This existed some 14 000 years ago and the gorge which it formed is now largely filled by glacial drift.

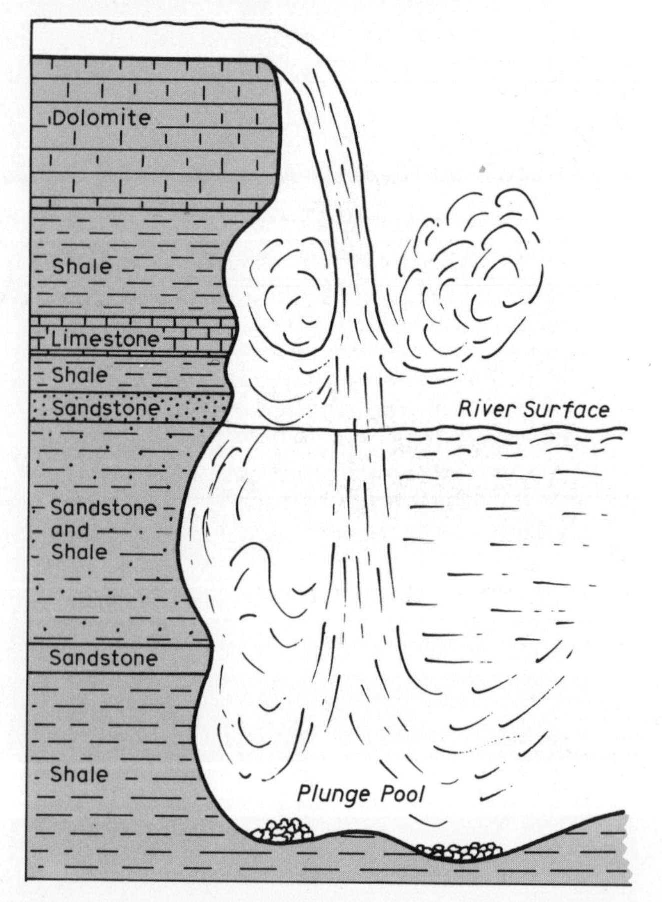

Fig 200 Section through Niagara Falls showing differential erosion of strata and formation of plunge pool

Most of the great waterfalls of the world, and probably a majority of the smaller ones, are variants of the **cap rock** type of which Niagara is a classic example. Near horizontal basalt flows are frequent producers of waterfalls as, for example, Victoria Falls on the Zambesi, or High Force on the Tees.

The overdeepening of valleys by glaciers frequently leaves tributary valleys 'hanging' above the main valley and this is another common origin for waterfalls—e.g. the Lodoor Falls on Watendlath Beck SE of Derwentwater, and the Bridal Veil Falls, Yosemite Valley, California.

Faulting, too, may result in waterfalls as in the case of the Kalambo Falls near the SE corner of Lake Tanganyika. Here a series of falls and cascades take the river from the plateau at 1500 m to the rift valley floor 750 m below.

Coastal erosion may, by removing the lower portion of a river with a steep gradient, produce a cliff over which a stream tumbles into the sea. Examples of such falls are to be found at Osmington Mills, a few miles E of Weymouth, and Hartland on the N coast of Devon, a short distance S of Hartland Point.

When the processes which bring about the recession of Niagara Falls are appreciated, it is not difficult to understand that all waterfalls, even the largest, are ephemeral features in the landscape. They are self-destructive, in the sense that the energy which they generate leads to their recession and to their eventual reduction until they are incorporated into the graded profile of the river.

FURTHER WORK

Practical

(1) Apply the descriptive and analytical method used for Niagara Falls in the preceding Study to any other

waterfalls for which photographs and/or large-scale maps are available.

(2) Better still, of course, visit a waterfall—even a very small one—and carefully observe such features as its height, volume of water, velocity above and below the falls (can you measure the velocity of the falling water?), evidence of recession, possible cause of origin, etc.

Reading

Anderson: *Splendour of Earth*, pp. 284–288 and 300–304
Gresswell: *Rivers and Valleys*, Ch. 7, pp. 72–80
——: *Physical Geography*, Ch. 17, pp. 187–190
Hardy & Monkhouse: *Physical Landscape in Pictures*, pp. 41–43
Holmes: *Principles of Physical Geology*, Ch. XVIII, pp. 508–509 and 522–527
Horrocks: *Physical Geography and Climatology*, Ch. 5, pp. 61–64
Monkhouse: *Principles of Physical Geography*, Ch. 6, pp. 116–120
——: *Landscape from the Air*, p. 20

POSTSCRIPT—STUDYING THE LANDSCAPE

The variety of landscapes is infinite, but whatever part of the Earth's surface is studied it is possible to divide the land into two categories—it is either sloping or it is flat. And this is true whether we are looking at the city street outside the school gates or a photograph of the Alps. Such a statement may seem so obvious that it is not worth making, but in fact if these two elements are isolated, described and analysed the student has progressed a long way in an understanding of Geomorphology.

SLOPES

Once any slope has been noticed, the steps in description must be:

(a) the steepness, i.e. the angle of slope or gradient;

(b) the form of the slope, i.e. convex, concave, straight, or a combination of these elements;

(c) the nature of the material forming the slope, e.g. bare rock, scree, grass with terracettes or without terracettes, or a man-made surface;

(d) the relation of the slope to nearby slopes, e.g. is it one side of a valley with a comparable facing slope?; is it the edge of a hill area whose other sides are bounded by steeper, less steep or similar slopes?

Having thus described our slope we can begin to think about its origins. No slope happens by chance; it must have a cause. Basically we can assign this cause either to a constructive process which builds up the slope or a destructive process which forms a slope by removing material (*Fig 202*).

Constructive processes include mountain building (folding and faulting) volcanic activity, morainic deposition by ice, dune formation by wind, river deposition (e.g. levées) and so on. Destructive processes include mainly the processes of erosion—by rivers, ice, the sea.

It may be that in a few cases one can attribute the slope almost entirely to structural factors. The case of bare rock surfaces parallel to the dip of the rock would be a case in point (*Fig 201*).

Fig 201 Marstrand, 20 km N of Göteborg

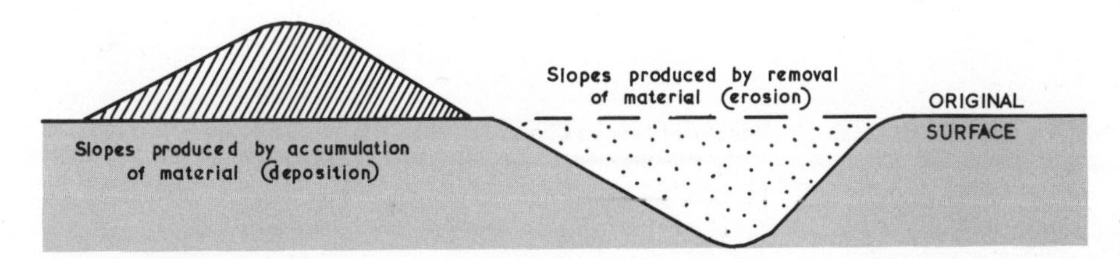

Fig 202 Depositional and erosional slopes

More usually, although one may see some relationship between slope and structure, it will clearly be necessary to invoke another process also and so the destructive elements of landscape formation must be considered. (The Giggleswick Scar studied in E.4 provides an example of a structural slope modified by erosive processes.)

We have seen that in the case of rivers the erosive effect is a combination of active erosion by the river, mass movement bringing weathered products to the river and transport by the river to remove the products of both erosion and mass movement. The form and nature of the slope may suggest the relative importance of these elements. *Study A2* has indicated evidence for active soil creep. Where mass movement provides the river at the foot of a slope with more material than it can remove, the surplus accumulating at the foot of the slope is likely to reduce the overall steepness. Conversely, where river erosion occurs more rapidly than weathering processes can keep up with, the slope is likely to be steepened.

If this is the case, which circumstances might be expected to produce concave, convex and straight slopes? Can you support this by observation?

Similar lines of argument can, of course, be applied to slopes produced by glacial erosion or to the features of sea cliffs.

To return to the slope which is the subject of study. It may now be possible to suggest, by reference to the form of the slope and the nature of its covering whether, at the present time, it is being actively steepened by erosion, degraded by deposition at its foot or in a stable state. In other words it would, in such circumstances, be possible to explain the slope in terms of processes operating at the present day.

However it may well be that neither structural factors nor present processes explain the form and nature of the observed slope. For instance the valley sides in a riverless chalk valley cannot be explained by the structure (perhaps horizontal chalk strata) nor is there any mechanism existing in the valley today which could remove vast quantities of rock debris to carve the valley. Similarly the steep slopes of the valley sides in the Cairngorms cannot be explained by the existing small rivers. Present day features suggest accumulation of material in the valley floor, yet the carving of the valleys involved the removal of enormous volumes of rock.

In such cases the student must say to himself, 'Structure offers no explanation; the present processes do not (fully) explain the observed facts—therefore the explanation must lie with a set of conditions which have operated in the past'. Such a line of thought should lead to the realization that the chalk valleys must once have had surface rivers, and that the Cairngorm slopes can only be explained in terms of a past glacial epoch. Having reached this stage the student may then say: 'I must seek other evidence for chalkland rivers (e.g. a dendritic pattern of valleys) or for glaciation (e.g. investigation of the long profile of the valley or, in the field, search for striations or morainic deposits).'

FLATS

The same line of reasoning can be applied to observed flat surfaces:

(i) Is the flatness related to structure, e.g. horizontal beds of resistant strata?

If not

(ii) Can the flat surface be explained in terms of processes at present operative, e.g. the flood plain of a river; a wave cut platform; the floor of a playa lake basin?

If not

(iii) Can past conditions be postulated which would offer an explanation, e.g. the floor of a lake long since disappeared (Lake Pickering); the remnant of a flood plain cut into by a rejuvenated river (Rakaia Gorge); the remnant of a level surface formed when sea level was considerably higher than it is today (Bodmin Moor)?

And finally

(iv) What other evidence must I seek on the map or in the field which will support my conclusion or force me to reject it and start the reasoning process again?

This mode of thinking represents *real* Geomorphology—the study of the shape of the earth's surface.

INDEX

ACKNOWLEDGEMENTS

We wish to thank the following for permission to use copyright photographs and material:

Cover Aerofilms Ltd.
Figure 2 Author
5 Aerofilms Ltd.
7 Author
9 Author
10 Official U.S. Navy Photograph, U.S. Naval Research Laboratory V. P. Robey 1949
21 Author
24 Hulton Educational Publication Ltd. from Gresswell: *Rivers and Valleys*
26 Geological Survey and Museum, Crown copyright reserved
32 Author
37 Author
40 & 41 Redrawn from Strahler, *Physical Geography*, John Wiley & Sons Inc.
44 Redrawn from Strahler, *op. cit.*
45 Aerofilms Ltd.
63 Redrawn from Holmes, *Principles of Physical Geography*, Thomas Nelson & Sons
69 Redrawn from Gresswell, *Beaches and Coast Lines*, Hulton Educational Publications Ltd.
74 Author
75 Author
78 J. Allan Cash
80 Redrawn from Holmes, *op. cit.*
82 Aerofilms Ltd.
83 Author
90 Author
98 Author
99 Peter Beck

Line diagrams by K. J. Wass, University College, London
Field Sketches by Reproduction Drawings Ltd., Oxford

The specially prepared map extracts are reproduced with the kind permission of the following:

British: The Ordnance Survey, Crown copyright reserved
American: U.S. Department of the Interior: Geological Survey
French: Institut Géographique National

Figure 100 Author
101 Eric Kay
105 Aerofilms Ltd.
111 Redrawn from Holmes, *op. cit.*
113 Aerofilms Ltd.
114 Aerofilms Ltd.
119 Geological Survey and Museum, Crown copyright reserved
120 Geological Survey and Museum, Crown copyright reserved
123 Finnish Tourist Association, Helsinki
125 Paul Popper Ltd.
128 Aerofilms Ltd.
135 Geological Survey and Museum, Crown copyright reserved
140 Aerofilms Ltd.
143 Aerofilms Ltd.
144 Redrawn from Holmes, *op. cit.*
145 Redrawn from Holmes, *op. cit.*
146 H. Drewes, U.S. Geological Survey
147 J. K. Stacy, U.S. Geological Survey
152 H. S. Gale, U.S. Geological Survey
155 Aerofilms Ltd.
158 Aerofilms Ltd.
164 Author
166 Eric Kay
167 Eric Kay
170 Eric Kay
173 Roy J. Westlake
174 Roy J. Westlake
181 Aerofilms Ltd.
184 U.S. Geological Survey
185 U.S. Geological Survey
188 J. Allan Cash
190 Eric Kay
195 Redrawn from Holmes, *op. cit.*
196 J. Allan Cash
201 Don Hurley
page 44 Barnaby's Picture Library
45 Camera Press